# The Algebra of Mohammed ben Musa

TRANSLATED BY
FRIEDRICH AUGUST ROSEN

CAMBRIDGE
UNIVERSITY PRESS

CAMBRIDGE UNIVERSITY PRESS

Cambridge, New York, Melbourne, Madrid, Cape Town,
Singapore, São Paulo, Delhi, Mexico City

Published in the United States of America by Cambridge University Press, New York

www.cambridge.org
Information on this title: www.cambridge.org/9781108055079

© in this compilation Cambridge University Press 2013

This edition first published 1831
This digitally printed version 2013

ISBN 978-1-108-05507-9 Paperback

This book reproduces the text of the original edition. The content and language reflect
the beliefs, practices and terminology of their time, and have not been updated.

Cambridge University Press wishes to make clear that the book, unless originally published
by Cambridge, is not being republished by, in association or collaboration with, or
with the endorsement or approval of, the original publisher or its successors in title.

# THE

# ALGEBRA

OF

# MOHAMMED BEN MUSA.

# Oriental Translation Fund.

INSTITUTED 1828.

UNDER THE PATRONAGE OF

## HIS MOST GRACIOUS MAJESTY

## GEORGE THE FOURTH.

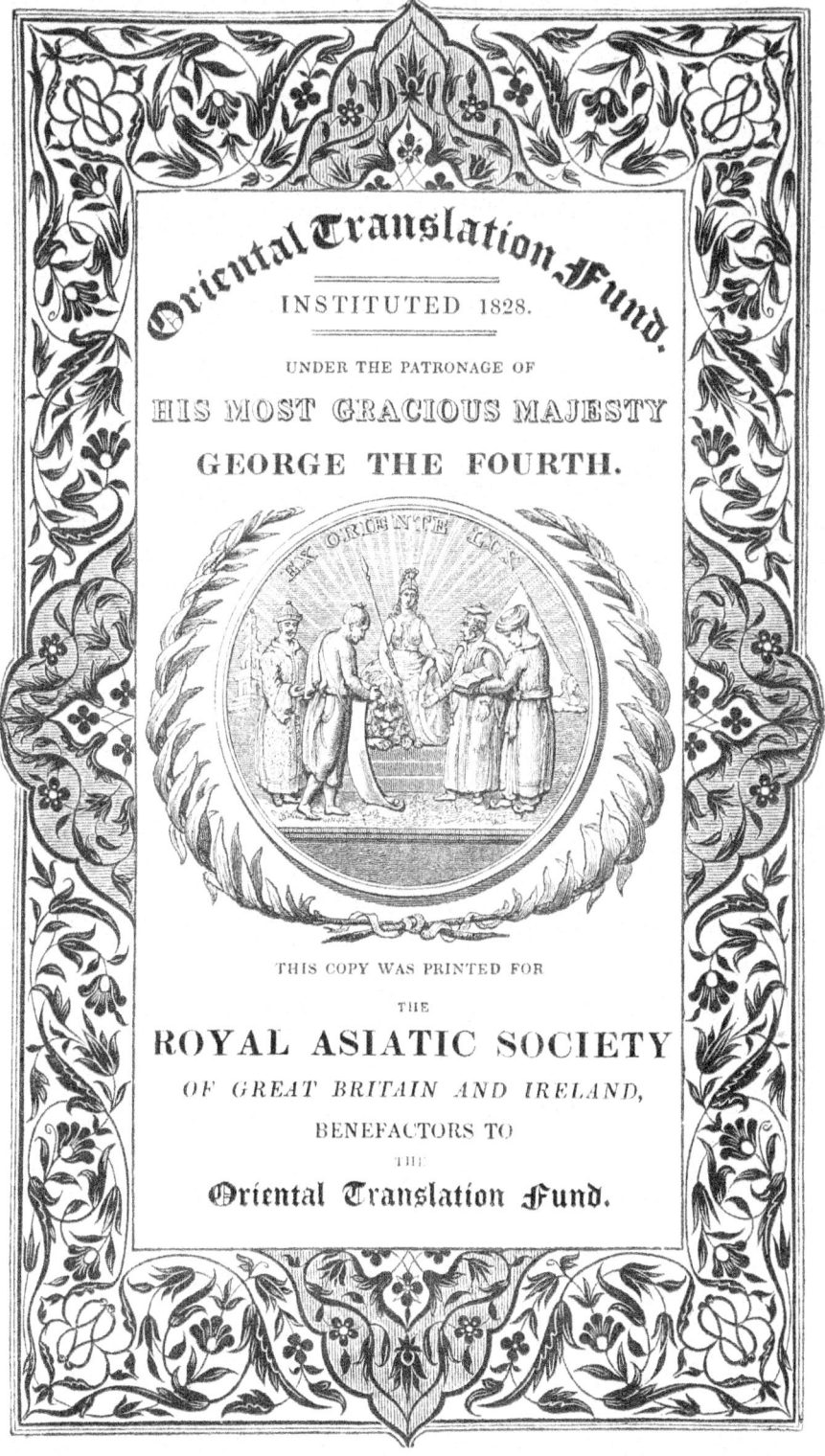

THIS COPY WAS PRINTED FOR

THE

## ROYAL ASIATIC SOCIETY

OF GREAT BRITAIN AND IRELAND,

BENEFACTORS TO

THE

Oriental Translation Fund.

# THE

# ALGEBRA

OF

# MOHAMMED BEN MUSA.

---

EDITED AND TRANSLATED

BY

FREDERIC ROSEN.

---

LONDON:
PRINTED FOR THE ORIENTAL TRANSLATION FUND:
AND SOLD BY
J. MURRAY, ALBEMARLE STREET;
PARBURY, ALLEN, & CO., LEADENHALL STREET;
THACKER & CO., CALCUTTA; TREUTTEL & WUERTZ, PARIS;
AND E. FLEISCHER, LEIPZIG.

1831.

PRINTED BY
J. L. COX, GREAT QUEEN STREET,
LONDON.

# PREFACE.

In the study of history, the attention of the observer is drawn by a peculiar charm towards those epochs, at which nations, after having secured their independence externally, strive to obtain an inward guarantee for their power, by acquiring eminence as great in science and in every art of peace as they have already attained in the field of war. Such an epoch was, in the history of the Arabs, that of the Caliphs AL MANSUR, HARUN AL RASHID, and AL MAMUN, the illustrious contemporaries of CHARLEMAGNE; to the glory of which era, in the volume now offered to the public, a new monument is endeavoured to be raised.

ABU ABDALLAH MOHAMMED BEN MUSA, of Khowarezm, who it appears, from his preface, wrote this Treatise at the command of the Caliph AL MAMUN, was for a long time considered as the original inventor of Algebra. "*Hæc ars olim a* MAHOMETE, MOSIS *Arabis filio, initium sumsit: etenim hujus rei locuples testis* LEO-

NARDUS PISANUS." Such are the words with which HIERONYMUS CARDANUS commences his *Ars Magna,* in which he frequently refers to the work here translated, in a manner to leave no doubt of its identity.

That he was not the inventor of the Art, is now well established; but that he was the first Mohammedan who wrote upon it, is to be found asserted in several Oriental writers. HAJI KHALFA, in his bibliographical work, cites the initial words of the treatise now before us,* and

---

\* I am indebted to the kindness of my friend Mr. GUSTAV FLUEGEL of Dresden, for a most interesting extract from this part of HAJI KHALFA'S work. Complete manuscript copies of the كشف الظنون are very scarce. The only two which I have hitherto had an opportunity of examining (the one bought in Egypt by Dr. EHRENBERG, and now deposited in the Royal Library at Berlin—the other among RICH's collection in the British Museum) are only abridgments of the original compilation, in which the quotation of the initial words of each work is generally omitted. The prospect of an edition and Latin translation of the complete original work, to be published by Mr. FLUEGEL, under the auspices of the Oriental Translation Committee, must under such circumstances be most gratifying to all friends of Asiatic literature.

states, in two distinct passages, that its author, MOHAMMED BEN MUSA, was the first Mussulman who had ever written on the solution of problems by the rules of completion and reduction. Two marginal notes in the Oxford manuscript—from which the text of the present edition is taken—and an anonymous Arabic writer, whose *Bibliotheca Philosophorum* is frequently quoted by CASIRI,* likewise maintain that this production of MOHAMMED BEN MUSA was the first work written on the subject† by a Mohammedan.

---

\* تاريخ الحكماء , written in the twelfth century. CASIRI *Bibliotheca Arabica Escurialensis*, T. I. 426. 428.

† The first of these marginal notes stands at the top of the first page of the manuscript, and reads thus : هذا اول كتاب وضع في الجبر والمقابلة في الاسلام ولهذا ذكر فيه من كل فن طرفا ليفيد الاصول في الجبر والمقابلة " This is the first book written on (the art of calculating by) completion and reduction by a Mohammedan: on this account the author has introduced into it rules of various kinds, in order to render useful the very rudiments of Algebra." The other scholium stands farther on: it is the same to which I have referred in my notes to the Arabic text, p. 177.

From the manner in which our author, in his preface, speaks of the task he had undertaken, we cannot infer that he claimed to be the inventor. He says that the Caliph AL MAMUN encouraged him to write a *popular* work on Algebra: an expression which would seem to imply that other treatises were then already extant. From a formula for finding the circumference of the circle, which occurs in the work itself (Text p. 51, Transl. p. 72), I have, in a note, drawn the conclusion, that part of the information comprised in this volume was derived from an Indian source; a conjecture which is supported by the direct assertion of the author of the *Bibliotheca Philosophorum* quoted by CASIRI (1.426, 428). That MOHAMMED BEN MUSA was conversant with Hindu science, is further evident from the fact\* that he abridged, at AL MAMUN's request—but before the accession of that prince to the caliphat—the *Sindhind*, or

---

\* Related by EBN AL ADAMI in the preface to his astronomical tables. CASIRI, I. 427, 428. COLEBROOKE, Dissertation, &c. p. lxiv. lxxii.

astronomical tables, translated by MOHAMMED BEN IBRAHIM AL FAZARI from the work of an Indian astronomer who visited the court of ALMANSUR in the 156th year of the Hejira (A.D. 773).

The science as taught by MOHAMMED BEN MUSA, in the treatise now before us, does not extend beyond quadratic equations, including problems with an affected square. These he solves by the same rules which are followed by DIOPHANTUS\*, and which are taught, though less comprehensively, by the Hindu mathematicians†. That he should have borrowed from DIOPHANTUS is not at all probable; for it does not appear that the Arabs had any knowledge of DIOPHANTUS' work before the middle of the fourth century after the Hejira, when ABU'L-WAFA BUZJANI rendered it into Arabic‡ It

---

\* See DIOPHANTUS, Introd. § 11. and Book iv. problems 32 and 33.

† *Lilavati*, p. 29, *Vijaganita*, p. 347, of Mr. COLEBROOKE's translation.

‡ CASIRI *Bibl. Arab. Escur.* I. 433. COLEBROOKE's Dissertation, &c. p. lxxii.

is far more probable that the Arabs received their first knowledge of Algebra from the Hindus, who furnished them with the decimal notation of numerals, and with various important points of mathematical and astronomical information.

But under whatever obligation our author may be to the Hindus, as to the subject matter of his performance, he seems to have been independent of them in the manner of digesting and treating it : at least the method which he follows in expounding his rules, as well as in showing their application, differs considerably from that of the Hindu mathematical writers. BHASKARA and BRAHMAGUPTA give dogmatical precepts, unsupported by argument, which, even by the metrical form in which they are expressed, seem to address themselves rather to the memory than to the reasoning faculty of the learner: MOHAMMED gives his rules in simple prose, and establishes their accuracy by geometrical illustrations. The Hindus give comparatively few examples, and are fond of investing the statement of their problems in rhetorical pomp: the Arab, on the contrary, is remarkably rich in examples, but he introduces them with the same perspicuous simplicity of style which distinguishes his rules. In solving their problems, the Hindus are satisfied with pointing at the result, and at the principal intermediate steps which lead to it: the Arab shows the working of each example at full length, keeping his view constantly fixed upon the two sides of the equation, as upon the two scales of a balance, and showing how any alteration in one side is counterpoised by a corresponding change in the other.

Besides the few facts which have already been mentioned in the course of this preface, little or nothing is known of our Author's life. He lived and wrote under the caliphat of AL MAMUN, and must therefore be distinguished from ABU JAFAR MOHAMMED BEN MUSA*,

---

* The father of the latter, MUSA BEN SHAKER, whose native country I do not find recorded, had been a robber or bandit in the earlier part of his life, but had afterwards found means to attach himself to the court of the Caliph AL-MAMUN; who, after MUSA's death, took care of

likewise a mathematician and astronomer, who flourished under the Caliph AL MOTADED (who reigned A.H. 279-289, A.D. 892-902).

---

the education of his three sons, MOHAMMED, AHMED, and AL HASSAN. (ABILFARAGII *Histor. Dyn.* p. 280. CASIRI, I. 386. 418). Each of the sons subsequently distinguished himself in mathematics and astronomy. We learn from ABULFARAJ (*l. c.* p. 281) and from EBN KHALLIKAN (art. ثابت بن قرة) that THABET BEN KORRAH, the well-known translator of the Almagest, was indebted to MOHAMMED for his introduction to AL MOTADED, and the men of science at the court of that caliph. EBN KHALLIKAN's words are:

فخرج من حران ونزل كفرتوثا واقام بها
مدة الي ان قدم محمد بن موسي من بلاد الروم راجعا
الي بغداد فاجتمع به فرأه فاضلا فصيحا فاستصحبه الي بغداد
وانزله في داره ووصله بالخليفة فادخله في جملة المنجمين

" (THABET BEN KORRAH) left Harran, and established himself at Kafratutha, where he remained till MOHAMMED BEN MUSA arrived there, on his return from the Greek dominions to Bagdad. The latter became acquainted with THABET and on seeing his skill and sagacity, invited THABET to accompany him to Bagdad, where MOHAMMED made him lodge at his own house, introduced him to the Caliph, and procured him an appointment in the body of astronomers." EBN KHALLIKAN here speaks of MOHAMMED BEN MUSA as of a well-known individual: he has however devoted no special article to an account of his life. It is possible

( xiii )

The manuscript from whence the text of the present edition is taken—and which is the only copy the existence of which I have as yet been able to trace—is preserved in the Bodleian collection at Oxford. It is, together with three other treatises on Arithmetic and Algebra, contained in the volume marked CMXVIII. *Hunt.* 214, *fol.*, and bears the date of the transcription A.H. 743 (A. D. 1342). It is written in a plain and legible hand, but unfortunately destitute of most of the diacritical points: a deficiency which has often been very sensibly felt; for though the nature of the subject matter can but seldom leave a doubt as to the general import of a sentence, yet the true reading of some passages, and the precise interpretation of others, remain involved in obscurity. Besides, there occur several omissions of words, and even of entire sentences; and also instances of words or short passages writ-

---

that the tour into the provinces of the Eastern Roman Empire here mentioned, was undertaken in search of some ancient Greek works on mathematics or astronomy.

ten twice over, or words foreign to the sense introduced into the text. In printing the Arabic part, I have included in brackets many of those words which I found in the manuscript, the genuineness of which I suspected, and also such as I inserted from my own conjecture, to supply an apparent hiatus.

The margin of the manuscript is partially filled with *scholia* in a very small and almost illegible character, a few specimens of which will be found in the notes appended to my translation. Some of them are marked as being extracted from a commentary (شرح) by AL MOZAIHAFI*, probably the same author, whose full name is JEMALEDDIN ABU ABDALLAH MOHAMMED BEN OMAR AL JAZA'I† AL MOZAIHAFI, and whose "Introduction to Arithmetic," (مقدمة في الحساب) is contained in the same volume with MOHAMMED's work in the Bodleian library.

Numerals are in the text of the work always

---

* Wherever I have met with this name, it is written without the diacritical points المرحفى, and my pronunciation rests on mere conjecture.

† الحراعى ( ? )

expressed by words: figures are only used in some of the diagrams, and in a few marginal notes.

The work had been only briefly mentioned in URIS' catalogue of the Bodleian manuscripts. Mr. H. T. COLEBROOKE first introduced it to more general notice, by inserting a full account of it, with an English translation of the directions for the solution of equations, simple and compound, into the notes of the "*Dissertation*" prefixed to his invaluable work, "*Algebra, with Arithmetic and Mensuration, from the Sanscrit of Brahmegupta and Bhascara.*" (London, 1817, 4to. pages lxxv-lxxix.)

The account of the work given by Mr. COLEBROOKE excited the attention of a highly distinguished friend of mathematical science, who encouraged me to undertake an edition and translation of the whole: and who has taken the kindest interest in the execution of my task. He has with great patience and care revised and corrected my translation, and has furnished the commentary, subjoined to the text, in the form of common algebraic notation. But my obligations to him are not confined to this only; for his luminous advice has enabled me to overcome many difficulties, which, to my own limited proficiency in mathematics, would have been almost insurmountable.

In some notes on the Arabic text which are appended to my translation, I have endeavoured, not so much to elucidate, as to point out for further enquiry, a few circumstances connected with the history of Algebra. The comparisons drawn between the Algebra of the Arabs and that of the early Italian writers might perhaps have been more numerous and more detailed; but my enquiry was here restricted by the want of some important works. MONTUCLA, COSSALI, HUTTON, and the Basil edition of CARDANUS' *Ars magna*, were the only sources which I had the opportunity of consulting.

# THE AUTHOR'S PREFACE.

IN THE NAME OF GOD, GRACIOUS AND MERCIFUL!

This work was written by MOHAMMED BEN MUSA, of KHOWAREZM. He commences it thus:

Praised be God for his bounty towards those who deserve it by their virtuous acts: in performing which, as by him prescribed to his adoring creatures, we express our thanks, and render ourselves worthy of the continuance (of his mercy), and preserve ourselves from change: acknowledging his might, bending before his power, and revering his greatness! He sent MOHAMMED (on whom may the blessing of God repose!) with the mission of a prophet, long after any messenger from above had appeared, when justice had fallen into neglect, and when the true way of life was sought for in vain. Through him he cured of blindness, and saved through him from perdition, and increased through him what before was small, and collected through him what before was scattered. Praised be God our Lord! and may his glory increase, and may all his names be hallowed—besides whom there is no God; and may his benediction rest on MOHAMMED the Prophet and on his descendants!

The learned in times which have passed away, and among nations which have ceased to exist, were constantly employed in writing books on the several departments of science and on the various branches of knowledge, bearing in mind those that were to come after them, and hoping for a reward proportionate to their ability, and trusting that their endeavours would meet with acknowledgment, attention, and remembrance—content as they were even with a small degree of praise; small, if compared with the pains which they had undergone, and the difficulties which they had encountered in revealing the secrets and obscurities of science.

(2) Some applied themselves to obtain information which was not known before them, and left it to posterity; others commented upon the difficulties in the works left by their predecessors, and defined the best method (of study), or rendered the access (to science) easier or

placed it more within reach; others again discovered mistakes in preceding works, and arranged that which was confused, or adjusted what was irregular, and corrected the faults of their fellow-labourers, without arrogance towards them, or taking pride in what they did themselves.

That fondness for science, by which God has distinguished the Imam al Mamun, the Commander of the Faithful (besides the caliphat which He has vouchsafed unto him by lawful succession, in the robe of which He has invested him, and with the honours of which He has adorned him), that affability and condescension which he shows to the learned, that promptitude with which he protects and supports them in the elucidation of obscurities and in the removal of difficulties,—has encouraged me to compose a short work on Calculating by (the rules of) Completion and Reduction, confining it to what is easiest and most useful in arithmetic, such as men constantly require in cases of inheritance, legacies, partition, law-suits, and trade, and in all their dealings with one another, or where the measuring of lands, the digging of canals, geometrical computation, and other objects of various sorts and kinds are concerned—relying on the goodness of my intention therein, and hoping that the learned will reward it, by obtaining (for me) through their prayers the excellence of the Divine mercy: in requital of which, may the choicest blessings and the abundant bounty of God be theirs! My confidence rests with God, in this as in every thing, and in Him I put my trust. He is the Lord of the Sublime Throne. May His blessing descend upon all the prophets and heavenly messengers!

# MOHAMMED BEN MUSA'S

## COMPENDIUM

### ON CALCULATING BY

# COMPLETION AND REDUCTION.

---

WHEN I considered what people generally want in calculating, I found that it always is a number.

I also observed that every number is composed of units, and that any number may be divided into units.

Moreover, I found that every number, which may be expressed from one to ten, surpasses the preceding by one unit: afterwards the ten is doubled or tripled, just as before the units were: thus arise twenty, thirty, &c., until a hundred; then the hundred is doubled and tripled in the same manner as the units and the tens, up to a thousand; then the thousand can be thus repeated at any complex number; and so forth to the utmost limit of numeration.

I observed that the numbers which are required in calculating by Completion and Reduction are of three kinds, namely, roots, squares, and simple numbers relative to neither root nor square.

A root is any quantity which is to be multiplied by itself, consisting of units, or numbers ascending, or fractions descending.*

A square is the whole amount of the root multiplied by itself.

A simple number is any number which may be pronounced without reference to root or square.

A number belonging to one of these three classes may be equal to a number of another class; you may say, for instance, " squares are equal to roots," or " squares are equal to numbers," or "roots are equal to numbers."†

(4) Of the case in which *squares are equal to roots*, this is an example. " A square is equal to five roots of the same ;"‡ the root of the square is five, and the square is twenty-five, which is equal to five times its root.

So you say, " one third of the square is equal to four roots;"§ then the whole square is equal to twelve roots; that is a hundred and forty-four; and its root is twelve.

Or you say, " five squares are equal to ten roots ;"∥ then one square is equal to two roots; the root of the square is two, and its square is four.

---

\* By the word root, is meant the simple power of the unknown quantity.

† $cx^2 = bx$      $cx^2 = a$      $bx = a$

‡ $x^2 = 5x$    $\therefore x = 5$

§ $\dfrac{x^2}{3} = 4x$    $\therefore x^2 = 12x$    $\therefore x = 12$

∥ $5x^2 = 10x$    $\therefore x^2 = 2x$    $\therefore x = 2$

In this manner, whether the squares be many or few, (*i. e.* multiplied or divided by any number), they are reduced to a single square; and the same is done with the roots, which are their equivalents; that is to say, they are reduced in the same proportion as the squares.

As to the case in which *squares are equal to numbers*; for instance, you say, " a square is equal to nine;"* then this is a square, and its root is three. Or " five squares are equal to eighty;"† then one square is equal to one-fifth of eighty, which is sixteen. Or " the half of the square is equal to eighteen;"‡ then the square is thirty-six, and its root is six.

Thus, all squares, multiples, and sub-multiples of them, are reduced to a single square. If there be only part of a square, you add thereto, until there is a whole square; you do the same with the equivalent in numbers.

As to the case in which *roots are equal to numbers;* for instance, " one root equals three in number;"§ then the root is three, and its square nine. Or " four roots are equal to twenty;"‖ then one root is equal to five, and the square to be formed of it is twenty-five. Or " half the root is equal to ten;"¶ then the

(5)

---

\* $x^2 = 9 \qquad x = 3$

† $5x^2 = 80 \therefore x^2 = \frac{80}{5} = 16$

‡ $\frac{x^2}{2} = 18 \therefore x^2 = 36 \therefore x = 6$

§ $x = 3$

‖ $4x = 20 \qquad \therefore x = 5$

¶ $\frac{x}{} = 10 \qquad \therefore x = 20$

whole root is equal to twenty, and the square which is formed of it is four hundred.

I found that these three kinds; namely, roots, squares, and numbers, may be combined together, and thus three compound species arise;* that is, " squares and roots equal to numbers;" " squares and numbers equal to roots;" "roots and numbers equal to squares."

*Roots and Squares are equal to Numbers;*† for instance, " one square, and ten roots of the same, amount to thirty-nine dirhems;" that is to say, what must be the square which, when increased by ten of its own roots, amounts to thirty-nine? The solution is this: you halve the number‡ of the roots, which in the present instance yields five. This you multiply by itself; the product is twenty-five. Add this to thirty-nine; the sum is sixty-four. Now take the root of this, which is eight, and subtract from it half the number of the roots, which is five; the remainder is three. This is the root of the square which you sought for; the square itself is nine.

---

\* The three cases considered are,
    1st. $cx^2 + bx = a$
    2d. $cx^2 + a = bx$
    3d. $cx^2 = bx + a$

† 1st case: $cx^2 + bx = a$
Example $x^2 + 10x = 39$
$x = \sqrt{[(\tfrac{1}{2})^2 + 39]} - \tfrac{10}{2}$
   $= \sqrt{64} - 5$
   $= 8 - 5 = 3$

‡ *i. e.* the coefficient.

The solution is the same when two squares or three, or more or less be specified;* you reduce them to one single square, and in the same proportion you reduce also the roots and simple numbers which are connected therewith.

For instance, " two squares and ten roots are equal to forty-eight dirhems;"† that is to say, what must be the amount of two squares which, when summed up and added to ten times the root of one of them, make up a sum of forty-eight dirhems? You must at first reduce the two squares to one; and you know that one square of the two is the moiety of both. Then reduce every thing mentioned in the statement to its half, and it will be the same as if the question had been, a square and five roots of the same are equal to twenty-four dirhems; or, what must be the amount of a square which, when added to five times its root, is equal to twenty-four dirhems? Now halve the number of the roots; the moiety is two and a half. Multiply that by itself; the product is six and a quarter. Add this to twenty-four; the sum is thirty dirhems and a quarter. Take the root of this; it is five and a half. Subtract from this the moiety of the number of the roots, that is two and a half; the

(6)

---

* $cx^2 + bx = a$ is to be reduced to the form $x^2 + \frac{b}{c}x = \frac{a}{c}$

† $2x^2 + 10x = 48$
$x^2 + 5x = 24$
$x = \sqrt{[(\frac{5}{2})^2 + 24]} - \frac{5}{2}$
$\phantom{x} = \sqrt{[6\frac{1}{4} + 24]} - 2\frac{1}{2}$
$\phantom{x} = \phantom{\sqrt{[}} 5\frac{1}{2} - 2\frac{1}{2} = 3$

remainder is three. This is the root of the square, and the square itself is nine.

The proceeding will be the same if the instance be, "half of a square and five roots are equal to twenty-eight dirhems;"* that is to say, what must be the amount of a square, the moiety of which, when added to the equivalent of five of its roots, is equal to twenty-eight dirhems? Your first business must be to complete your square, so that it amounts to one whole square. This you effect by doubling it. Therefore double it, and double also that which is added to it, as well as what is equal to it. Then you have a square and ten roots, equal to fifty-six dirhems. Now halve the roots; the moiety is five. Multiply this by itself; the product is twenty-five. Add this to fifty-six; the sum is eighty-one. Extract the root of this; it is nine. Subtract from this the moiety of the number of roots, which is five; the remainder is four. This is the root of the square which you sought for; the square is sixteen, and half the square eight.

(7) Proceed in this manner, whenever you meet with squares and roots that are equal to simple numbers: for it will always answer.

---

$$* \frac{x^2}{2} + 5x = 28$$
$$x^2 + 10x = 56$$
$$x = \sqrt{[(\tfrac{10}{2})^2 + 56]} - \tfrac{10}{2}$$
$$= \sqrt{25 + 56} - 5$$
$$= \sqrt{81} - 5$$
$$= 9 - 5 = 4$$

*Squares and Numbers are equal to Roots;*\* for instance, "a square and twenty-one in numbers are equal to ten roots of the same square." That is to say, what must be the amount of a square, which, when twenty-one dirhems are added to it, becomes equal to the equivalent of ten roots of that square? Solution: Halve the number of the roots; the moiety is five. Multiply this by itself; the product is twenty-five. Subtract from this the twenty-one which are connected with the square; the remainder is four. Extract its root; it is two. Subtract this from the moiety of the roots, which is five; the remainder is three. This is the root of the square which you required, and the square is nine. Or you may add the root to the moiety of the roots; the sum is seven; this is the root of the square which you sought for, and the square itself is forty-nine.

When you meet with an instance which refers you to this case, try its solution by addition, and if that do not serve, then subtraction certainly will. For in this case both addition and subtraction may be employed, which will not answer in any other of the three cases in which

---

\* 2d case. $cx^2 + a = bx$
Example. $x^2 + 21 = 10x$
$x = \frac{10}{2} \pm \sqrt{[(\frac{10}{2})^2 - 21]}$
$= 5 \pm \sqrt{25 - 21}$
$= 5 \pm \sqrt{4}$
$= 5 \pm 2$

( 12 )

the number of the roots must be halved. And know, that, when in a question belonging to this case you have halved the number of the roots and multiplied the moiety by itself, if the product be less than the number of dirhems connected with the square, then the instance is impossible;* but if the product be equal to (8) the dirhems by themselves, then the root of the square is equal to the moiety of the roots alone, without either addition or subtraction.

In every instance where you have two squares, or more or less, reduce them to one entire square, † as I have explained under the first case.

*Roots and Numbers are equal to Squares;*‡ for instance, " three roots and four of simple numbers are equal to a square." Solution: Halve the roots; the moiety is one and a half. Multiply this by itself; the product is two and a quarter. Add this to the four; the sum is

---

* If in an equation, of the form $x^2 + a = bx$, $(\frac{b}{2})^2 < a$, the case supposed in the equation cannot happen. If $(\frac{b}{2})^2 = a$, then $x = \frac{b}{2}$

† $cx^2 + a = bx$ is to be reduced to $x^2 + \frac{a}{c} = \frac{b}{c}x$

‡ 3d case $cx^2 = bx + a$
Example $x^2 = 3x + 4$
$$\begin{aligned}
x^2 &= \sqrt{[(\tfrac{3}{2})^2 + 4]} + \tfrac{3}{2} \\
&= \sqrt{(1\tfrac{1}{4})^2 + 4} + 1\tfrac{1}{2} \\
&= \sqrt{2\tfrac{1}{4} + 4} + 1\tfrac{1}{2} \\
&= \sqrt{6\tfrac{1}{4}} + 1\tfrac{1}{2} \\
&= 2\tfrac{1}{2} + 1\tfrac{1}{2} = 4
\end{aligned}$$

six and a quarter. Extract its root; it is two and a half. Add this to the moiety of the roots, which was one and a half; the sum is four. This is the root of the square, and the square is sixteen.

Whenever you meet with a multiple or sub-multiple of a square, reduce it to one entire square.

These are the six cases which I mentioned in the introduction to this book. They have now been explained. I have shown that three among them do not require that the roots be halved, and I have taught how they must be resolved. As for the other three, in which halving the roots is necessary, I think it expedient, more accurately, to explain them by separate chapters, in which a figure will be given for each case, to point out the reasons for halving.

*Demonstration of the Case:* " *a Square and ten Roots are equal to thirty-nine Dirhems.*"\*

The figure to explain this a quadrate, the sides of which are unknown. It represents the square, the which, or the root of which, you wish to know. This is the figure A B, each side of which may be considered as one of its roots; and if you multiply one of these sides by any number, then the amount of that number may be looked upon as the number of the roots which are added to the square. Each side of the quadrate represents the root of the square; and, as in the instance,

---

\* Geometrical illustration of the case, $x^2 + 10x = 39$

the roots were connected with the square, we may take one-fourth of ten, that is to say, two and a half, and combine it with each of the four sides of the figure. Thus with the original quadrate A B, four new parallelograms are combined, each having a side of the quadrate as its length, and the number of two and a half as its breadth; they are the parallelograms C, G, T, and K. We have now a quadrate of equal, though unknown sides; but in each of the four corners of which a square piece of two and a half multiplied by two and a half is wanting. In order to compensate for this want and to complete the quadrate, we must add (to that which we have already) four times the square of two and a half, that is, twenty-five. We know (by the statement) that the first figure, namely, the quadrate representing the square, together with the four parallelograms around it, which represent the ten roots, is equal to thirty-nine of numbers. If to this we add twenty-five, which is the equivalent of the four quadrates at the corners of the figure A B, by which the great figure D H is completed, then we know that this together makes sixty-four. One side of this great quadrate is its root, that is, eight. If we subtract twice a fourth of ten, that is five, from eight, as from the two extremities of the side of the great quadrate D H, then the remainder of such a side will be three, and that is the root of the square, or the side of the original figure A B. It must be observed, that we have halved the number of the roots, and added the product of the moiety multiplied by itself to the number thirty-nine, in order to complete the great figure in its (10) four corners; because the fourth of any number multiplied by itself, and then by four, is equal to the product of the moiety of that number multiplied by itself.* Accordingly, we multiplied only the moiety of the roots by itself, instead of multiplying its fourth by itself, and then by four. This is the figure:

The same may also be explained by another figure. We proceed from the quadrate A B, which represents the square. It is our next business to add to it the ten roots of the same. We halve for this purpose the ten, so that it becomes five, and construct two quadrangles on two sides of the quadrate A B, namely, G and D, the length of each of them being five, as the moiety of the ten roots, whilst the breadth of each is equal to a side of the quadrate A B. Then a quadrate remains opposite the corner of the quadrate A B. This is equal to five multiplied by five: this five being half of the number of the roots which we have added to each of the two sides of the first quadrate. Thus we know that

---

* $4 \times \left(\frac{b}{4}\right)^2 = \left(\frac{b}{2}\right)^2$

( 16 )

(11) the first quadrate, which is the square, and the two quadrangles on its sides, which are the ten roots, make together thirty-nine. In order to complete the great quadrate, there wants only a square of five multiplied by five, or twenty-five. This we add to thirty-nine, in order to complete the great square S H. The sum is sixty-four. We extract its root, eight, which is one of the sides of the great quadrangle. By subtracting from this the same quantity which we have before added, namely five, we obtain three as the remainder. This is the side of the quadrangle A B, which represents the square; it is the root of this square, and the square itself is nine. This is the figure:—

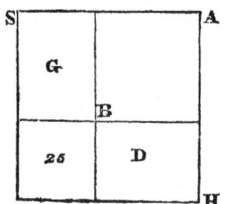

*Demonstration of the Case:* " *a Square and twenty-one Dirhems are equal to ten Roots.*"\*

We represent the square by a quadrate A D, the length of whose side we do not know. To this we join a parallelogram, the breadth of which is equal to one of the sides of the quadrate A D, such as the side H N. This paralellogram is H B. The length of the two

---

\* Geometrical illustration of the case, $x^2 + 21 = 10x$

figures together is equal to the line H C. We know that its length is ten of numbers; for every quadrate has equal sides and angles, and one of its sides multiplied by a unit is the root of the quadrate, or multiplied by two it is twice the root of the same. As it is stated, therefore, that a square and twenty-one of numbers are equal to ten roots, we may conclude that the length of the line H C is equal to ten of numbers, since the line C D represents the root of the square. We now divide the line C H into two equal parts at the point G: the line G C is then equal to H G. It is also evident that (12) the line G T is equal to the line C D. At present we add to the line G T, in the same direction, a piece equal to the difference between C G and G T, in order to complete the square. Then the line T K becomes equal to K M, and we have a new quadrate of equal sides and angles, namely, the quadrate M T. We know that the line T K is five; this is consequently the length also of the other sides: the quadrate itself is twenty-five, this being the product of the multiplication of half the number of the roots by themselves, for five times five is twenty-five. We have perceived that the quadrangle H B represents the twenty-one of numbers which were added to the quadrate. We have then cut off a piece from the quadrangle H B by the line K T (which is one of the sides of the quadrate M T), so that only the part T A remains. At present we take from the line K M the piece K L, which is equal to G K; it then appears that the line T G is equal to M L; more-

D

over, the line K L, which has been cut off from K M, is equal to K G; consequently, the quadrangle MR is equal to T A. Thus it is evident that the quadrangle H T, augmented by the quadrangle M R, is equal to the quadrangle H B, which represents the twenty-one. The whole quadrate M T was found to be equal to twenty-five. If we now subtract from this quadrate, M T, the quadrangles H T and M R, which are equal to twenty-one, there remains a small quadrate K R, which represents the difference between twenty-five and twenty-one. This is four; and its root, represented by the line R G, which is equal to G A, is two. If you (13) subtract this number two from the line C G, which is the moiety of the roots, then the remainder is the line A C; that is to say, three, which is the root of the original square. But if you add the number two to the line C G, which is the moiety of the number of the roots, then the sum is seven, represented by the line C R, which is the root to a larger square. However, if you add twenty-one to this square, then the sum will likewise be equal to ten roots of the same square. Here is the figure:—

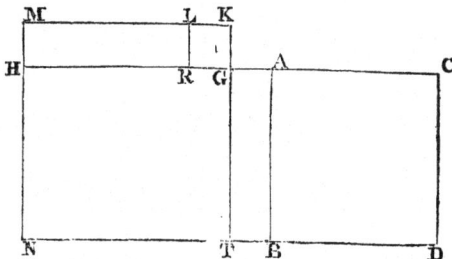

*Demonstration of the Case: "three Roots and four of Simple Numbers are equal to a Square."\**

Let the square be represented by a quadrangle, the sides of which are unknown to us, though they are equal among themselves, as also the angles. This is the quadrate A D, which comprises the three roots and the four of numbers mentioned in this instance. In every quadrate one of its sides, multiplied by a unit, is its root. We now cut off the quadrangle H D from the quadrate A D, and take one of its sides H C for three, which is the number of the roots. The same is equal to R D. It follows, then, that the quadrangle H B represents the four of numbers which are added to the roots. Now we halve the side C H, which is equal to three roots, at the point G; from this division we construct the square H T, which is the product of half the roots (or one and a half) multiplied by themselves, that is to say, two and a quarter. We add then to the line G T a piece equal to the line A H, namely, the piece T L; accordingly the line G L becomes equal to A G, and the line K N equal to T L. Thus a new quadrangle, with equal sides and angles, arises, namely, the quadrangle G M; and we find that the line A G is equal to M L, and the same line A G is equal to G L. By these means the line C G remains equal to N R, and the line M N equal to T L, and from the quadrangle H B a piece equal to the quadrangle K L is cut off.

(14)

---

\* Geometrical illustration of the 3d case, $x^2 = 3x + 4$

But we know that the quadrangle A R represents the four of numbers which are added to the three roots. The quadrangle A N and the quadrangle K L are together equal to the quadrangle A R, which represents the four of numbers.

We have seen, also, that the quadrangle G M comprises the product of the moiety of the roots, or of one and a half, multiplied by itself; that is to say two and a quarter, together with the four of numbers, which are represented by the quadrangles A N and K L. There remains now from the side of the great original quadrate A D, which represents the whole square, only the moiety of the roots, that is to say, one and a half, namely, the line G C. If we add this to the line A G, which is the root of the quadrate G M, being equal to two and a half; then this, together with C G, or the moiety of the three roots, namely, one and a half, makes four, which is the line A C, or the root to a square, which is represented by the quadrate A D. Here follows the figure. This it was which we were desirous to explain.

(15)

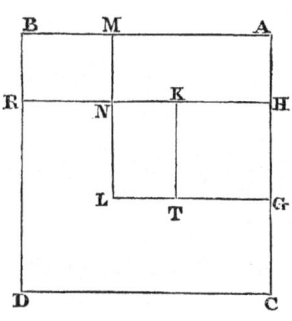

( 21 )

We have observed that every question which requires equation or reduction for its solution, will refer you to one of the six cases which I have proposed in this book. I have now also explained their arguments. Bear them, therefore, in mind.

## ON MULTIPLICATION.

I shall now teach you how to multiply the unknown numbers, that is to say, the roots, one by the other, if they stand alone, or if numbers are added to them, or if numbers are subtracted from them, or if they are subtracted from numbers; also how to add them one to the other, or how to subtract one from the other.

Whenever one number is to be multiplied by another, the one must be repeated as many times as the other contains units.*

If there are greater numbers combined with units to be added to or subtracted from them, then four multiplications are necessary;† namely, the greater numbers by the greater numbers, the greater numbers by the

---

* If $x$ is to be multiplied by $y$, $x$ is to be repeated as many times as there are units in $y$.

† If $x \pm a$ is to be multiplied by $y \pm b$, $x$ is to be multiplied by $y$, $x$ is to be multiplied by $b$, $a$ is to be multiplied by $y$, and $a$ is to be multiplied by $b$.

units, the units by the greater numbers, and the units by the units.

If the units, combined with the greater numbers, are positive, then the last multiplication is positive; if they are both negative, then the fourth multiplication is likewise positive. But if one of them is positive, and one (16) negative, then the fourth multiplication is negative.*

For instance, " ten and one to be multiplied by ten and two."† Ten times ten is a hundred; once ten is ten positive; twice ten is twenty positive, and once two is two positive; this altogether makes a hundred and thirty-two.

But if the instance is " ten less one, to be multiplied by ten less one,"‡ then ten times ten is a hundred; the

---

\* In multiplying $(x \pm a)$ by $(y \pm b)$

$$+a \times +b = +ab$$
$$-a \times -b = +ab$$
$$+a \times -b = -ab$$
$$-a \times +b = -ab$$

† $(10+1) \times (10+2)$
$= 10 \times 10 \ldots\ldots 100$
$+\phantom{0}1 \times 10 \ldots\ldots \phantom{0}10$
$+\phantom{0}2 \times 10 \ldots\ldots \phantom{0}20$
$+\phantom{0}1 \times \phantom{0}2 \ldots\ldots \phantom{00}2$
$\phantom{+0000000000}+132$

‡ $(10-1)(10-1)$
$= 10 \times \phantom{0}10 .. +100$
$-\phantom{0}1 \times \phantom{0}10 .. -\phantom{0}10$
$-\phantom{0}1 \times \phantom{0}10 .. -\phantom{0}10$
$-\phantom{0}1 \times -1 .. +\phantom{00}1$
$\phantom{+000000000}+\phantom{0}81$

negative one by ten is ten negative; the other negative one by ten is likewise ten negative, so that it becomes eighty: but the negative one by the negative one is one positive, and this makes the result eighty-one.

Or if the instance be "ten and two, to be multipled by ten less one,"* then ten times ten is a hundred, and the negative one by ten is ten negative; the positive two by ten is twenty positive; this together is a hundred and ten; the positive two by the negative one gives two negative. This makes the product a hundred and eight.

I have explained this, that it might serve as an introduction to the multiplication of unknown sums, when numbers are added to them, or when numbers are subtracted from them, or when they are subtracted from numbers.

For instance: "Ten less thing (the signification of thing being root) to be multipled by ten."† You begin by taking ten times ten, which is a hundred; less thing by ten is ten roots negative; the product is therefore a hundred less ten things.

---

$$* \ (10+2) \times (10-1) =$$
$$10 \times 10 \ldots \ldots \ 100$$
$$- \ \ 1 \times 10 \ldots \ldots \ -10$$
$$+10 \times \ 2 \ldots \ldots \ +20$$
$$- \ \ 1 \times \ 2 \ldots \ldots \ - \ 2$$
$$\overline{\phantom{xxxxxxxx}108}$$

† $(10-x) \times 10 = 10 \times 10 - 10x = 100 - 10x.$

If the instance be: "ten and thing to be multiplied by ten,"* then you take ten times ten, which is a hundred, and thing by ten is ten things positive; so that the product is a hundred plus ten things.

(17) If the instance be: "ten and thing to be multiplied by itself,"† then ten times ten is a hundred, and ten times thing is ten things; and again, ten times thing is ten things; and thing multiplied by thing is a square positive, so that the whole product is a hundred dirhems and twenty things and one positive square.

If the instance be: "ten minus thing to be multiplied by ten minus thing,"‡ then ten times ten is a hundred; and minus thing by ten is minus ten things; and again, minus thing by ten is minus ten things. But minus thing multiplied by minus thing is a positive square. The product is therefore a hundred and a square, minus twenty things.

In like manner if the following question be proposed to you: "one dirhem minus one-sixth to be multiplied by one dirhem minus one-sixth;"§ that is to say, five-sixths by themselves, the product is five and twenty parts of a dirhem, which is divided into six and thirty parts, or two-thirds and one-sixth of a sixth. Computation: You multiply one dirhem by one dirhem, the

---

*$(10+x) \times 10 = 10 \times 10 + 10x = 100 + 10x$
†$(10+x)(10+x) = 10 \times 10 + 10x + 10x + x^2 = 100 + 20x + x^2$
‡$(10-x) \times (10-x) = 10 \times 10 - 10x - 10x + x^2 = 100 - 20x + x^2$
§$(1-\frac{1}{6}) \times (1-\frac{1}{6}) = 1 - \frac{1}{3} + \frac{1}{6} \times \frac{1}{6} = \frac{2}{3} + \frac{1}{6} \times \frac{1}{6}$; i.e. $\frac{25}{36} = \frac{2}{3} + \frac{1}{6} \times \frac{1}{6}$

product is one dirhem; then one dirhem by minus one-sixth, that is one-sixth negative; then, again, one dirhem by minus one-sixth is one-sixth negative: so far, then, the result is two-thirds of a dirhem: but there is still minus one-sixth to be multiplied by minus one-sixth, which is one-sixth of a sixth positive; the product is, therefore, two-thirds and one sixth of a sixth.

If the instance be, " ten minus thing to be multiplied by ten and thing," then you say,* ten times ten is a hundred; and minus thing by ten is ten things negative; and thing by ten is ten things positive; and minus thing by thing is a square positive; therefore, the product is a hundred dirhems, minus a square.

If the instance be, " ten minus thing to be multiplied by thing,"† then you say, ten multiplied by thing is ten things; and minus thing by thing is a square negative; therefore, the product is ten things minus a square. (18)

If the instance be, " ten and thing to be multiplied by thing less ten,"‡ then you say, thing multiplied by ten is ten things positive; and thing by thing is a square positive; and minus ten by ten is a hundred dirhems negative; and minus ten by thing is ten things negative. You say, therefore, a square minus a hundred dirhems; for, having made the reduction, that is to say, having removed the ten things positive by the ten things

---

\* $(10-x)(10+x) = 10 \times 10 - 10x + 10x - x^2 = 100 - x^2$

† $(10-x) \times x = 10x - x^2$

‡ $(10+x)(x-10) = 10x + x^2 - 100 - 10x = x^2 - 100$

negative, there remains a square minus a hundred dirhems.

If the instance be, "ten dirhems and half a thing to be multiplied by half a dirhem, minus five things,"* then you say, half a dirhem by ten is five dirhems positive; and half a dirhem by half a thing is a quarter of thing positive; and minus five things by ten dirhems is fifty roots negative. This altogether makes five dirhems minus forty-nine things and three quarters of thing. After this you multiply five roots negative by half a root positive: it is two squares and a half negative. Therefore, the product is five dirhems, minus two squares and a half, minus forty-nine roots and three quarters of a root.

If the instance be, "ten and thing to be multiplied by thing less ten,"† then this is the same as if it were said thing and ten by thing less ten. You say, therefore, thing multiplied by thing is a square positive; and ten by thing is ten things positive; and minus ten by thing is ten things negative. You now remove the positive by the negative, then there only remains a square. Minus ten multiplied by ten is a hundred, to be subtracted from the square. This, therefore, altogether, is a square less a hundred dirhems.

(19) Whenever a positive and a negative factor concur in

---

*$(10+\frac{x}{2})(\frac{1}{2}-5x) = \frac{10}{2}+\frac{x}{4}-50x-\frac{5}{2}x^2 = 5-49\frac{3}{4}x-2\frac{1}{2}x^2$

†$(10+x)(x-10) = (x+10)(x-10) = x^2+10x-10x-100 = x^2-100$

a multiplication, such as thing positive and minus thing, the last multiplication gives always the negative product. Keep this in memory.

## ON ADDITION AND SUBTRACTION.

Know that the root of two hundred minus ten, added to twenty minus the root of two hundred, is just ten.*

The root of two hundred, minus ten, subtracted from twenty minus the root of two hundred, is thirty minus twice the root of two hundred; twice the root of two hundred is equal to the root of eight hundred.†

A hundred and a square minus twenty roots, added to fifty and ten roots minus two squares,‡ is a hundred and fifty, minus a square and minus ten roots.

A hundred and a square, minus twenty roots, diminished by fifty and ten roots minus two squares, is fifty dirhems and three squares minus thirty roots.§

I shall hereafter explain to you the reason of this by a figure, which will be annexed to this chapter.

If you require to double the root of any known or unknown square, (the meaning of its duplication being

---

\* $20 - \sqrt{200} + (\sqrt{200} - 10) = 10$

† $20 - \sqrt{200} - (\sqrt{200} - 10) = 30 - 2\sqrt{200} = 30 - \sqrt{800}$

‡ $50 + 10x - 2x^2 + (100 + x^2 - 20x) = 150 - 10x - x^2$

§ $100 + x^2 - 20x - [50 - 2x^2 + 10x] = 50 + 3x^2 - 30x$

that you multiply it by two) then it will suffice to multiply two by two, and then by the square;* the root of the product is equal to twice the root of the original square.

If you require to take it thrice, you multiply three by three, and then by the square; the root of the product is thrice the root of the original square.

Compute in this manner every multiplication of the roots, whether the multiplication be more or less than two.†

(20) If you require to find the moiety of the root of the square, you need only multiply a half by a half, which is a quarter; and then this by the square: the root of the product will be half the root of the first square.‡

Follow the same rule when you seek for a third, or a quarter of a root, or any larger or smaller quota§ of it, whatever may be the denominator or the numerator.

*Examples of this:* If you require to double the root of nine,‖ you multiply two by two, and then by nine: this gives thirty six; take the root of this, it is six, and this is double the root of nine.

---

* $2\sqrt{x^2} = \sqrt{4x^2}$
$3\sqrt{x^2} = \sqrt{9x^2}$
† $n\sqrt{x^2} = \sqrt{n^2 x^2}$
‡ $\frac{1}{2}\sqrt{x^2} = \sqrt{\frac{x^2}{4}}$
§ $\frac{1}{n}\sqrt{x^2} = \sqrt{\frac{x^2}{n^2}}$
‖ $2\sqrt{9} = \sqrt{4 \times 9} = \sqrt{36} = 6$

In the same manner, if you require to triple the root of nine,* you multiply three by three, and then by nine: the product is eighty-one; take its root, it is nine, which becomes equal to thrice the root of nine.

If you require to have the moiety of the root of nine,† you multiply a half by a half, which gives a quarter, and then this by nine; the result is two and a quarter: take its root; it is one and a half, which is the moiety of the root of nine.

You proceed in this manner with every root, whether positive or negative, and whether known or unknown.

## ON DIVISION.

If you will divide the root of nine by the root of four,‡ you begin with dividing nine by four, which gives two and a quarter: the root of this is the number which you require—it is one and a half.

If you will divide the root of four by the root of nine,§ you divide four by nine; it is four-ninths of the unit: the root of this is two divided by three; namely, two-thirds of the unit.

---

\* $3\sqrt{9} = \sqrt{9 \times 9} = \sqrt{81} = 9$

† $\frac{1}{2}\sqrt{9} = \sqrt{\frac{9}{4}} = \sqrt{2\frac{1}{4}} = 1\frac{1}{2}$

‡ $\frac{\sqrt{9}}{\sqrt{4}} = \sqrt{\frac{9}{4}} = \sqrt{2\frac{1}{4}} = 1\frac{1}{2}$

§ $\frac{\sqrt{4}}{\sqrt{9}} = \sqrt{\frac{4}{9}} = \frac{2}{3}$

If you wish to divide twice the root of nine by the root of four, or of any other square*, you double the (21) root of nine in the manner above shown to you in the chapter on Multiplication, and you divide the product by four, or by any number whatever. You perform this in the way above pointed out.

In like manner, if you wish to divide three roots of nine, or more, or one-half or any multiple or sub-multiple of the root of nine, the rule is always the same:† follow it, the result will be right.

If you wish to multiply the root of nine by the root of four,‡ multiply nine by four; this gives thirty-six; take its root, it is six; this is the root of nine, multiplied by the root of four.

Thus, if you wish to multiply the root of five by the root of ten,§ multiply five by ten: the root of the product is what you have required.

If you wish to multiply the root of one-third by the root of a half,∥ you multiply one-third by a half: it is one-sixth: the root of one-sixth is equal to the root of one-third, multiplied by the root of a half.

If you require to multiply twice the root of nine by

---

$$* \; \frac{2\sqrt{9}}{\sqrt{4}} = \sqrt{\frac{36}{4}} = \sqrt{9} = 3$$

$$\dagger \; \frac{m\sqrt{p^2}}{\sqrt{q^2}} = \sqrt{\frac{m^2 p^2}{q^2}}$$

$$\ddagger \; \sqrt{4} \times \sqrt{9} = \sqrt{4 \times 9} = \sqrt{36} = 6$$

$$\S \; \sqrt{10} \times \sqrt{5} = \sqrt{5 \times 10} = \sqrt{50}$$

$$\| \; \sqrt{\tfrac{1}{2}} \times \sqrt{\tfrac{1}{3}} = \sqrt{\tfrac{1}{2} \times \tfrac{1}{3}} = \sqrt{\tfrac{1}{6}}$$

thrice the root of four,\* then take twice the root of nine, according to the rule above given, so that you may know the root of what square it is. You do the same with respect to the three roots of four in order to know what must be the square of such a root. You then multiply these two squares, the one by the other, and the root of the product is equal to twice the root of nine, multiplied by thrice the root of four.

You proceed in this manner with all positive or negative roots.

*Demonstrations.* (22)

The argument for the root of two hundred, minus ten, added to twenty, minus the root of two hundred, may be elucidated by a figure:

Let the line A B represent the root of two hundred; let the part from A to the point C be the ten, then the remainder of the root of two hundred will correspond to the remainder of the line A B, namely to the line C B. Draw now from the point B a line to the point D, to represent twenty; let it, therefore, be twice as long as the line A C, which represents ten; and mark a part of it from the point B to the point H, to be equal to the line A B, which represents the root of two hundred; then the remainder of the twenty will be equal to the part of the line, from the point H to the point D. As

---

\* $3\sqrt{4} \times 2\sqrt{9} = \sqrt{9 \times 4} \times \sqrt{4 \times 9} = \sqrt{36 \times 36} = 36$

our object was to add the remainder of the root of two hundred, after the subtraction of ten, that is to say, the line C B, to the line H D, or to twenty, minus the root of two hundred, we cut off from the line B H a piece equal to C B, namely, the line S H. We know already that the line A B, or the root of two hundred, is equal to the line B H, and that the line A C, which represents the ten, is equal to the line S B, as also that the remainder of the line A B, namely, the line C B is equal to the remainder of the line B H, namely, to S H. Let us add, therefore, this piece S H, to the line H D. We have already seen that from the line B D, or twenty, a piece equal to A C, which is ten, was cut off, namely, the piece B S. There remains after this the line S D, which, consequently, is equal to ten. This it was that we intended to elucidate. Here follows the figure.

(23)

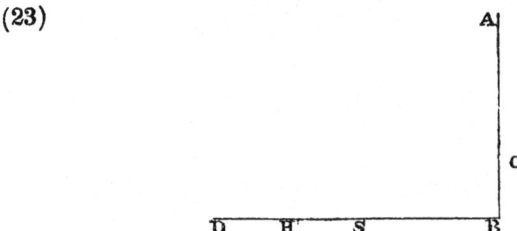

The argument for the root of two hundred, minus ten, to be subtracted from twenty, minus the root of two hundred, is as follows. Let the line A B represent the root of two hundred, and let the part thereof, from A to the point C, signify the ten mentioned in the instance. We draw now from the point B, a line towards the point D, to signify twenty. Then we trace from B to the

( 33 )

point H, the same length as the length of the line which represents the root of two hundred; that is of the line A B. We have seen that the line C B is the remainder from the twenty, after the root of two hundred has been subtracted. It is our purpose, therefore, to subtract the line C B from the line H D; and we now draw from the point B, a line towards the point S, equal in length to the line A C, which represents the ten. Then the whole line S D is equal to S B, plus B D, and we perceive that all this added together amounts to thirty. We now cut off from the line H D, a piece equal to C B, namely, the line H G; thus we find that the line G D is the remainder from the line S D, which signifies thirty. We see also that the line B H is the root of two hundred and that the line S B and B C is likewise the root of two hundred. Now the line H G is equal to C B; therefore the piece subtracted from the line S D, which represents thirty, is equal to twice the root of two hundred, or once the root of eight hundred. (24) This it is that we wished to elucidate.

Here follows the figure:

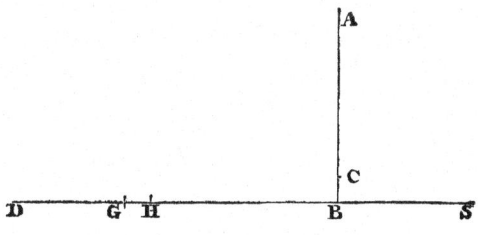

As for the hundred and square minus twenty roots added to fifty, and ten roots minus two squares, this does

F

not admit of any figure, because there are three different species, *viz.* squares, and roots, and numbers, and nothing corresponding to them by which they might be represented. We had, indeed, contrived to construct a figure also for this case, but it was not sufficiently clear.

The elucidation by words is very easy. You know that you have a hundred and a square, minus twenty roots. When you add to this fifty and ten roots, it becomes a hundred and fifty and a square, minus ten roots. The reason for these ten negative roots is, that from the twenty negative roots ten positive roots were subtracted by reduction. This being done, there remains a hundred and fifty and a square, minus ten roots. With the hundred a square is connected. If you subtract from this hundred and square the two squares negative connected with fifty, then one square disappears by reason of the other, and the remainder is a hundred and fifty, minus a square, and minus ten roots.

This it was that we wished to explain.

## OF THE SIX PROBLEMS.

BEFORE the chapters on computation and the several (25) species thereof, I shall now introduce six problems, as instances of the six cases treated of in the beginning of this work. I have shown that three among these cases, in order to be solved, do not require that the roots be halved, and I have also mentioned that the calculating by completion and reduction must always necessarily lead you to one of these cases. I now subjoin these problems, which will serve to bring the subject nearer to the understanding, to render its comprehension easier, and to make the arguments more perspicuous.

### *First Problem.*

I have divided ten into two portions; I have multiplied the one of the two portions by the other; after this I have multiplied the one of the two by itself, and the product of the multiplication by itself is four times as much as that of one of the portions by the other.*

Computation: Suppose one of the portions to be thing, and the other ten minus thing: you multiply

---

\* $x^2 = 4x(10-x) = 40x - 4x^2$
$5x^2 = 40x$
$x^2 = 8x$
$x = 8$; $(10-x) = 2$

thing by ten minus thing; it is ten things minus a square. Then multiply it by four, because the instance states "four times as much." The result will be four times the product of one of the parts multiplied by the other. This is forty things minus four squares. After this you multiply thing by thing, that is to say, one of the portions by itself. This is a square, which is equal to forty things minus four squares. Reduce it now by the four squares, and add them to the one square. Then the equation is: forty things are equal to five squares; and one square will be equal to eight roots, that is, sixty-four; the root of this is eight, and this is one of the two portions, namely, that which is to
(26) be multiplied by itself. The remainder from the ten is two, and that is the other portion. Thus the question leads you to one of the six cases, namely, that of "squares equal to roots." Remark this.

*Second Problem.*

I have divided ten into two portions: I have multiplied each of the parts by itself, and afterwards ten by itself: the product of ten by itself is equal to one of the two parts multiplied by itself, and afterwards by two and seven-ninths; or equal to the other multiplied by itself, and afterwards by six and one-fourth.*

---

* $10^2 = x^2 \times 2\frac{7}{9}$
$100 = x^2 \times \frac{25}{9}$
$\frac{9}{25} \times 100 = x^2$
$36 = x^2$
$6 = x$

Computation : Suppose one of the parts to be thing, and the other ten minus thing. You multiply thing by itself, it is a square; then by two and seven-ninths, this makes it two squares and seven-ninths of a square. You afterwards multiply ten by ten; it is a hundred, which much be equal to two squares and seven-ninths of a square. Reduce it to one square, through division by nine twenty-fifths;* this being its fifth and four-fifths of its fifth, take now also the fifth and four-fifths of the fifth of a hundred; this is thirty-six, which is equal to one square. Take its root, it is six. This is one of the two portions; and accordingly the other is four. This question leads you, therefore, to one of the six cases, namely, " squares equal to numbers."

### Third Problem.

I have divided ten into two parts. I have afterwards divided the one by the other, and the quotient was four.†

Computation : Suppose one of the two parts to be (27) thing, the other ten minus thing. Then you divide ten minus thing by thing, in order that four may be obtained. You know that if you multiply the quotient by the divisor, the sum which was divided is restored.

---

* $\frac{9}{25} = \frac{1}{5} \times \frac{4}{5} + \frac{1}{5}$

† $\frac{10-x}{x} = 4$

$10 - x = 4x$

$10 = 5x$

$2 = x$

( 38 )

In the present question the quotient is four and the divisor is thing. Multiply, therefore, four by thing; the result is four things, which are equal to the sum to be divided, which was ten minus thing. You now reduce it by thing, which you add to the four things. Then we have five things equal to ten; therefore one thing is equal to two, and this is one of the two portions. This question refers you to one of the six cases, namely, "roots equal to numbers."

### Fourth Problem.

I have multiplied one-third of thing and one dirhem by one-fourth of thing and one dirhem, and the product was twenty.*

Computation: You multiply one-third of thing by one-fourth of thing; it is one-half of a sixth of a square. Further, you multiply one dirhem by one-third of thing, it is one-third of thing; and one dirhem by one-fourth of thing, it is one-fourth of thing; and one dirhem by one dirhem, it is one dirhem. The result of this is: the moiety of one-sixth of a square, and one-third of thing, and one-fourth of thing, and one dirhem, is equal to twenty dirhems. Subtract now the one dirhem from

---

\* $(\frac{1}{3}x+1)(\frac{1}{4}x+1) = 20$
$\frac{x^2}{12} + \frac{1}{3}x + \frac{1}{4}x + 1 = 20$
$\frac{x^2}{12} + \frac{7}{12}x = 19$
$x^2 + 7x = 228$
$x = \sqrt{\frac{49}{4} + 228} - \frac{7}{2} = 12$

these twenty dirhems, there remain nineteen dirhems, equal to the moiety of one-sixth of a square, and one-third of thing, and one-fourth of thing. Now make your square a whole one: you perform this by multiplying all that you have by twelve. Thus you have one square and seven roots, equal to two hundred and twenty-eight dirhems. Halve the number of the roots, and multiply it by itself; it is twelve and one-fourth. Add this to the numbers, that is, to two hundred and twenty-eight; (28) the sum is two hundred and forty and one quarter. Extract the root of this; it is fifteen and a half. Subtract from this the moiety of the roots, that is, three and a half, there remains twelve, which is the square required. This question leads you to one of the cases, namely, "squares and roots equal to numbers."

### Fifth Problem.

I have divided ten into two parts; I have then multiplied each of them by itself, and when I had added the products together, the sum was fifty-eight dirhems.*

Computation: Suppose one of the two parts to be thing, and the other ten minus thing. Multiply ten minus thing by itself; it is a hundred and a square minus twenty things. Then multiply thing by thing; it

---

* $x^2 + (10-x)^2 = 58$
$2x^2 - 20x + 100 = 58$
$x^2 - 10x + 50 = 29$
$x^2 + 21 = 10x$
$x = 5 \pm \sqrt{25 - 21} = 5 \pm 2 = 7$ or $3$

is a square. Add both together. The sum is a hundred, plus two squares minus twenty things, which are equal to fifty-eight dirhems. Take now the twenty negative things from the hundred and the two squares, and add them to fifty-eight; then a hundred, plus two squares, are equal to fifty-eight dirhems and twenty things. Reduce this to one square, by taking the moiety of all you have. It is then: fifty dirhems and a square, which are equal to twenty-nine dirhems and ten things. Then reduce this, by taking twenty-nine from fifty; there remains twenty-one and a square, equal to ten things. Halve the number of the roots, it is five; multiply this by itself, it is twenty-five; take from this the twenty-one which are connected with the square, the remainder

(29) is four. Extract the root, it is two. Subtract this from the moiety of the roots, namely, from five, there remains three. This is one of the portions; the other is seven. This question refers you to one of the six cases, namely " squares and numbers equal to roots."

### Sixth Problem.

I have multiplied one-third of a root by one-fourth of a root, and the product is equal to the root and twenty-four dirhems.*

---

$$* \; \frac{x}{3} \times \frac{x}{4} = x + 24$$
$$\frac{x^2}{12} = x + 24$$
$$x^2 = 12x + 288$$
$$x = 6 + \sqrt{36 + 288} = 6 + 18 = 24$$

( 41 )

Computation: Call the root thing; then one-third of thing is multiplied by one-fourth of thing; this is the moiety of one-sixth of the square, and is equal to thing and twenty-four dirhems. Multiply this moiety of one-sixth of the square by twelve, in order to make your square a whole one, and multiply also the thing by twelve, which yields twelve things; and also four-and-twenty by twelve: the product of the whole will be two hundred and eighty-eight dirhems and twelve roots, which are equal to one square. The moiety of the roots is six. Multiply this by itself, and add it to two hundred and eighty-eight, it will be three hundred and twenty-four. Extract the root from this, it is eighteen; add this to the moiety of the roots, which was six; the sum is twenty-four, and this is the square sought for. This question refers you to one of the six cases, namely, "roots and numbers equal to squares."

## VARIOUS QUESTIONS.

If a person puts such a question to you as: "I have (30) divided ten into two parts, and multiplying one of these by the other, the result was twenty-one;"* then

---

\* $(10-x)x = 21$
$10x - x^2 = 21$
which is to be reduced to
$x^2 + 21 = 10x$
$x = 5 \pm \sqrt{25 - 21} = 5 \pm 2$

G

you know that one of the two parts is thing, and the other ten minus thing. Multiply, therefore, thing by ten minus thing; then you have ten things minus a square, which is equal to twenty-one. Separate the square from the ten things, and add it to the twenty-one. Then you have ten things, which are equal to twenty-one dirhems and a square. Take away the moiety of the roots, and multiply the remaining five by itself; it is twenty-five. Subtract from this the twenty-one which are connected with the square; the remainder is four. Extract its root, it is two. Subtract this from the moiety of the roots, namely, five; there remain three, which is one of the two parts. Or, if you please, you may add the root of four to the moiety of the roots; the sum is seven, which is likewise one of the parts. This is one of the problems which may be resolved by addition and subtraction.

If the question be: "I have divided ten into two parts, and having multiplied each part by itself, I have subtracted the smaller from the greater, and the remainder was forty;"* then the computation is—you multiply ten minus thing by itself, it is a hundred plus one square minus twenty things; and you also multiply thing by

---

* $(10-x)^2 - x^2 = 40$
$100 - 20x = 40$
$100 = 20x + 40$
$60 = 20x$
$3 = x$

thing, it is one square. Subtract this from a hundred and a square minus twenty things, and you have a hundred, minus twenty things, equal to forty dirhems. Separate now the twenty things from a hundred, and add them to the forty; then you have a hundred, equal to twenty things and forty dirhems. Subtract now forty from a hundred; there remains sixty dirhems, equal to twenty things: therefore one thing is equal to three, which is one of the two parts.

If the question be: "I have divided ten into two parts, and having multiplied each part by itself, I have put them together, and have added to them the difference of the two parts previously to their multiplication, and the amount of all this is fifty-four;"* then the computation is this: You multiply ten minus thing by itself; it is a hundred and a square minus twenty things. Then multiply also the other thing of the ten by itself; it is one square. Add this together, it will be a hundred plus two squares minus twenty things. It was stated that the difference of the two parts before multiplication should be added to them. You say, therefore, the difference between them is ten minus two things.

---

* $(10-x)^2 + x^2 + (10-x) - x = 54$
$100 - 20x + 2x^2 + 10 - 2x = 54$
$100 - 22x + 2x^2 = 54$
$55 - 11x + x^2 = 27$
$x^2 + 28 = 11x$
$x = \frac{11}{2} \pm \sqrt{\frac{121}{4} - 28} = \frac{11 \pm 3}{2} = 7 \text{ or } 4$

The result is a hundred and ten and two squares minus twenty-two things, which are equal to fifty-four dirhems. Having reduced and equalized this, you may say, a hundred and ten dirhems and two squares are equal to fifty-four dirhems and twenty-two things. Reduce now the two squares to one square, by taking the moiety of all you have. Thus it becomes fifty-five dirhems and a square, equal to twenty-seven dirhems and eleven things. Subtract twenty-seven from fifty-five, there remain (32) twenty-eight dirhems and a square, equal to eleven things. Halve now the things, it will be five and a half; multiply this by itself, it is thirty and a quarter. Subtract from it the twenty-eight which are combined with the square, the remainder is two and a fourth. Extract its root, it is one and a half. Subtract this from the moiety of the roots, there remain four, which is one of the two parts.

If one say, "I have divided ten into two parts; and have divided the first by the second, and the second by the first, and the sum of the quotient is two dirhems and one-sixth;"* then the computation is this: If you multiply each part by itself, and add the products together, then their sum is equal to one of the parts

---

\* $\dfrac{10-x}{x} + \dfrac{x}{10-x} = 2\tfrac{1}{6}$

$100 + 2x^2 - 20x = x(10-x) \times 2\tfrac{1}{6} = 21\tfrac{2}{3}x - 2\tfrac{1}{6}x^2$

$100 + 4\tfrac{1}{6}x^2 = 41\tfrac{2}{3}x$

$24 + x^2 = 10x$

$x = 5 \pm \sqrt{25-24} = 5 \pm 1 = 4 \text{ or } 6$

( 45 )

multiplied by the other, and again by the quotient which is two and one-sixth. Multiply, therefore, ten less thing by itself; it is a hundred and a square less ten things. Multiply thing by thing; it is one square. Add this together; the sum is a hundred plus two squares less twenty things, which is equal to thing multiplied by ten less thing; that is, to ten things less a square, multiplied by the sum of the quotients arising from the division of the two parts, namely, two and one-sixth. We have, therefore, twenty-one things and two-thirds of thing less two squares and one-sixth, equal to a hundred plus two squares less twenty things. Reduce this by adding the two squares and one-sixth to a hundred plus two squares less twenty things, and add the twenty negative things from the hundred plus the two squares to the twenty-one things and two-thirds of thing. Then you have a hundred plus four squares (33) and one-sixth of a square, equal to forty-one things and two-thirds of thing. Now reduce this to one square. You know that one square is obtained from four squares and one-sixth, by taking a fifth and one-fifth of a fifth.* Take, therefore, the fifth and one-fifth of a fifth of all that you have. Then it is twenty-four and a square, equal to ten roots; because ten is one-fifth and one-fifth of the fifth of the forty-one things and two-thirds of a thing. Now halve the roots; it gives five. Multiply this

---

* $4\frac{1}{6} = \frac{25}{6}$, and $\frac{6}{25} = \frac{1}{5} + \frac{1}{5} \times \frac{1}{5}$

by itself; it is five and-twenty. Subtract from this the twenty-four, which are connected with the square; the remainder is one. Extract its root; it is one. Subtract this from the moiety of the roots, which is five. There remains four, which is one of the two parts.

Observe that, in every case, where any two quantities whatsoever are divided, the first by the second and the second by the first, if you multiply the quotient of the one division by that of the other, the product is always one.*

If some one say: "You divide ten into two parts; multiply one of the two parts by five, and divide it by the other: then take the moiety of the quotient, and add this to the product of the one part, multiplied by five; the sum is fifty dirhems;"† then the computation is this: Take thing, and multiply it by five. This is now to be divided by the remainder of the ten, that is, by ten less thing; and of the quotient the moiety is to be taken.

(34) You know that if you divide five things by ten less thing, and take the moiety of the quotient, the result is

---

\* $\dfrac{a}{b} \times \dfrac{b}{a} = 1$

† $\dfrac{5x}{2(10-x)} + 5x = 50$

$\dfrac{x}{2(10-x)} + x = 10$

$x^2 + 100 = 20\frac{1}{2}x$

$x = \frac{41}{4} - \frac{9}{4} = 8$

the same as if you divide the moiety of five things by ten less thing. Take, therefore, the moiety of five things; it is two things and a half: and this you require to divide by ten less thing. Now these two things and a half, divided by ten less thing, give a quotient which is equal to fifty less five things: for the question states: add this (the quotient) to the one part multiplied by five, the sum will be fifty. You have already observed, that if the quotient, or the result of the division, be multiplied by the divisor, the dividend, or capital to be divided, is restored. Now, your capital, in the present instance, is two things and a half. Multiply, therefore, ten less thing by fifty less five things. Then you have five hundred dirhems and five squares less a hundred things, which are equal to two things and a half. Reduce this to one square. Then it becomes a hundred dirhems and a square less twenty things, equal to the moiety of thing. Separate now the twenty things from the hundred dirhems and square, and add them to the half thing. Then you have a hundred dirhems and a square, equal to twenty things and a half. Now halve the things, multiply the moiety by itself, subtract from this the hundred, extract the root of the remainder, and subtract this from the moiety of the roots, which is ten and one-fourth: the remainder is eight; and this is one of the portions.

If some one say: "You divide ten into two parts: multiply the one by itself; it will be equal to the other

( 48 )

taken eighty-one times."*  Computation: You say, ten less thing, multiplied by itself, is a hundred plus a square less twenty things, and this is equal to eighty-one things. Separate the twenty things from a hundred and a square, and add them to eighty-one. It will then be a hundred plus a square, which is equal to a hundred and one roots. Halve the roots; the moiety is fifty and a half. Multiply this by itself, it is two thousand five hundred and fifty and a quarter. Subtract from this one hundred; the remainder is two thousand four hundred and fifty and a quarter. Extract the root from this; it is forty-nine and a half. Subtract this from the moiety of the roots, which is fifty and a half. There remains one, and this is one of the two parts.

(35)

If some one say : " I have purchased two measures of wheat or barley, each of them at a certain price. I afterwards added the expences, and the sum was equal to the difference of the two prices, added to the difference of the measures."†

---

* $(10-x)^2 = 81x$
$100 - 20x + x^2 = 81x$
$x^2 + 100 = 101x$
$x = \frac{101}{2} - \sqrt{\frac{101^2}{4} - 100} = 50\frac{1}{2} - 49\frac{1}{2} = 1$

† The purchaser does not make a clear enunciation of the terms of his bargain. He intends to say, "I bought $m$ bushels of wheat, and $n$ bushels of barley, and the wheat was $r$ times dearer than the barley. The sum I expended was equal to the difference in the quantities, added to the difference in the prices of the grain."

Computation: Take what numbers you please, for it is indifferent; for instance, four and six. Then you say: I have bought each measure of the four for thing; and accordingly you multiply four by thing, which gives four things; and I have bought the six, each for the moiety of thing, for which I have bought the four; or, if you please, for one-third, or one-fourth, or for any other quota of that price, for it is indifferent. Suppose that you have bought the six measures for the moiety of thing, then you multiply the moiety of thing by six; this gives three things. Add them to the four things; the sum is seven things, which must be equal to the difference of the two quantities, which is two measures, plus the difference of the two prices, which is a moiety of thing. You have, therefore, seven things, equal to two and a moiety of thing. Remove, now, this moiety of thing, by subtracting it from the seven things. There remain six things and a half, equal to two dir- (36) hems: consequently, one thing is equal to four-thirteenths of a dirhem. The six measures were bought, each at one-half of thing; that is, at two-thirteenths of a dirhem. Accordingly, the expenses amount to eight-and-twenty thirteenths of a dirhem, and this sum is equal to the difference of the two quantities; namely,

---

If $x$ is the price of the barley, $rx$ is the price of the wheat; whence, $mrx + nx = (m - n) + (rx - x)$; $\therefore x = \dfrac{m-n}{mr+n+r-1}$ and the sum expended is $\dfrac{(mr+n) \times (m-n)}{mr+n+r-1}$.

the two measures, the arithmetical equivalent for which is six-and-twenty thirteenths, added to the difference of the two prices, which is two-thirteenths: both differences together being likewise equal to twenty-eight parts.

If he say: "There are two numbers,* the difference of which is two dirhems. I have divided the smaller by the larger, and the quotient was just half a dirhem."† Suppose one of the two numbers* to be thing, and the other to be thing plus two dirhems. By the division of thing by thing plus two dirhems, half a dirhem appears as quotient. You have already observed, that by multiplying the quotient by the divisor, the capital which you divided is restored. This capital, in the present case, is thing. Multiply, therefore, thing and two dirhems by half a dirhem, which is the quotient; the product is half one thing plus one dirhem; this is equal to thing. Remove, now, half a thing on account

---

* In the original, "squares." The word square is used in the text to signify either, 1st, a square, properly so called, fractional or integral; 2d, a rational integer, not being a square number; 3d, a rational fraction, not being a square; 4th, a quadratic surd, fractional or integral.

$$† \frac{x}{x+2} = \frac{1}{2}$$
$$x = \frac{x+2}{2} = \frac{x}{2} + 1$$
$$\frac{x}{2} = 1 \text{ and } x + 2 = 4$$

of the other half thing; there remains one dirhem, equal to half a thing. Double it, then you have one thing, equal to two dirhems. Consequently, the other number* is four.

If some one say: "I have divided ten into two parts; I have multiplied the one by ten and the other by itself, and the products were the same;"† then the computation is this: You multiply thing by ten; it is ten things. Then multiply ten less thing by itself; it is a hundred (37) and a square less twenty things, which is equal to ten things. Reduce this according to the rules, which I have above explained to you.

In like manner, if he say: "I have divided ten into two parts; I have multiplied one of the two by the other, and have then divided the product by the difference of the two parts before their multiplication, and the result of this division is five and one-fourth;"‡ the computation will be this: You subtract thing from ten; there remain ten less thing. Multiply the one by the other, it is ten things less a square. This is the product of the multiplication of one of the two parts by the other. At

---

\* "Square" in the original.

† $10x = (10-x)^2 = 100 - 20x + x^2$
$x = 15 - \sqrt{225 - 100} = 15 - \sqrt{125}$

‡ $\dfrac{x(10-x)}{10-2x} = 5\tfrac{1}{4}$
$10x - x^2 = 52\tfrac{1}{2} - 10\tfrac{1}{2}x$
$20\tfrac{1}{2}x = x^2 + 52\tfrac{1}{2}$
$x = 10\tfrac{1}{4} - 7\tfrac{1}{4} = 3$

present you divide this by the difference between the two parts, which is ten less two things. The quotient of this division is, according to the statement, five and a fourth. If, therefore, you muliply five and one-fourth by ten less two things, the product must be equal to the above amount, obtained by multiplication, namely, ten things less one square. Multiply now five and one-fourth by ten less two squares. The result is fifty-two dirhems and a half less ten roots and a half, which is equal to ten roots less a square. Separate now the ten roots and a half from the fifty-two dirhems, and add them to the ten roots less a square; at the same time separate this square from them, and add it to the fifty-two dirhems and a half. Thus you find twenty roots and a half, equal to fifty-two dirhems and a half and one square. Now continue reducing it, conformably to the rules explained at the commencement of this book.

(38) If the question be: "There is a square, two-thirds of one-fifth of which are equal to one-seventh of its root;" then the square is equal to one root and half a seventh of a root; and the root consists of fourteen-fifteenths of the square.* The computation is this: You

---

\* $\frac{2}{3} \times \frac{1}{5} x^2 = \frac{x}{7}$

$x^2 = 7\frac{1}{2} \times \frac{x}{7} = 1\frac{1}{14} x$

$x = 1\frac{1}{14}$

$x^2 = 1\frac{29}{196}$

$\frac{2}{15} x^2 = \frac{30}{196} \times \frac{x}{7}$

multiply two-thirds of one-fifth of the square by seven and a half, in order that the square may be completed. Multiply that which you have already, namely, one-seventh of its root, by the same. The result will be, that the square is equal to one root and half a seventh of the root; and the root of the square is one and a half seventh; and the square is one and twenty-nine one hundred and ninety-sixths of a dirhem. Two-thirds of the fifth of this are thirty parts of the hundred and ninety-six parts. One-seventh of its root is likewise thirty parts of a hundred and ninety-six.

If the instance be: "Three-fourths of the fifth of a square are equal to four-fifths of its root,"* then the computation is this: You add one-fifth to the four-fifths, in order to complete the root. This is then equal to three and three-fourths of twenty parts, that is, to fifteen eightieths of the square. Divide now eighty by fifteen; the quotient is five and one-third. This is the root of the square, and the square is twenty-eight and four-ninths.

If some one say: "What is the amount of a square-root,† which, when multiplied by four times itself,

---

* $\frac{3}{4} \times \frac{1}{5}x^2 = \frac{4}{5}x$

$\frac{3\frac{3}{4}}{20}x$, or $\frac{15}{80}x$, or $\frac{3}{16}x = 1$

$\therefore x = \frac{16}{3} = 5\frac{1}{3}$

† "Square" in the original.

amounts to twenty?*" the answer is this: If you multiply it by itself it will be five: it is therefore the root of five.

If somebody ask you for the amount of a square-root,† which when multiplied by its third amounts to ten,‡ the solution is, that when multiplied by itself it will amount to thirty; and it is consequently the root of thirty.

(39) If the question be: "To find a quantity†, which when multiplied by four times itself, gives one-third of the first quantity as product,"§ the solution is, that if you multiply it by twelve times itself, the quantity itself must re-appear: it is the moiety of one moiety of one-third.

If the question be: "A square, which when multiplied by its root gives three times the original square as product,"‖ then the solution is: that if you multiply the root by one-third of the square, the original square is

---

\* $4x^2 = 20$
$x = \sqrt{5}$

† " Square " in the original.

‡ $x \times \frac{x}{3} = 10$
$x^2 = 30$
$x = \sqrt{30}$

§ $x \times 4x = \frac{x}{3}$
$x = \frac{1}{12}$

‖ $x^2 \times x = 3x^2$
$x = 3$

restored; its root must consequently be three, and the square itself nine.

If the instance be: "To find a square, four roots of which, multiplied by three roots, restore the square with a surplus of forty-four dirhems,"* then the solution is: that you multiply four roots by three roots, which gives twelve squares, equal to a square and forty-four dirhems. Remove now one square of the twelve on account of the one square connected with the forty-four dirhems. There remain eleven squares, equal to forty-four dirhems. Make the division, the result will be four, and this is the square.

If the instance be: "A square, four of the roots of which multiplied by five of its roots produce twice the square, with a surplus of thirty-six dirhems;"† then the solution is: that you multiply four roots by five roots, which gives twenty squares, equal to two squares and thirty-six dirhems. Remove two squares from the twenty on account of the other two. The remainder is eighteen squares, equal to thirty-six dirhems. Divide now thirty-six dirhems by eighteen; the quotient is two, and this is the square.

---

$$* \quad 4x \times 3x = x^2 + 44$$
$$11x^2 = 44$$
$$x^2 = 4$$
$$x = 2$$
$$† \quad 4x \times 5x = 2x^2 + 36$$
$$18x^2 = 36$$
$$x^2 = 2$$

(40) In the same manner, if the question be: "A square, multiply its root by four of its roots, and the product will be three times the square, with a surplus of fifty dirhems."† Computation: You multiply the root by four roots, it is four squares, which are equal to three squares and fifty dirhems. Remove three squares from the four; there remains one square, equal to fifty dirhems. One root of fifty, multiplied by four roots of the same, gives two hundred, which is equal to three times the square, and a residue of fifty dirhems.

If the instance be: "A square, which when added to twenty dirhems, is equal to twelve of its roots,"† then the solution is this: You say, one square and twenty dirhems are equal to twelve roots. Halve the roots and multiply them by themselves; this gives thirty-six. Subtract from this the twenty dirhems, extract the root from the remainder, and subtract it from the moiety of the roots, which is six. The remainder is the root of the square: it is two dirhems, and the square is four.

If the instance be: "To find a square, of which if one-third be added to three dirhems, and the sum be subtracted from the square, the remainder multiplied by

---

$$* \quad 4x^2 = 3x^2 + 50$$
$$x^2 = 50$$
$$† \quad x^2 + 20 = 12x$$
$$x = 6 \pm \sqrt{36-20} = 6 \pm 4 = 10 \text{ or } 2$$

itself restores the square;"* then the computation is this: If you subtract one-third and three dirhems from the square, there remain two-thirds of it less three dirhems. This is the root. Multiply therefore two-thirds of thing less three dirhems by itself. You say two-thirds by two-thirds is four ninths of a square; and less two-thirds by three dirhems is two roots: and again, two-thirds by three dirhems is two roots; and less three dirhems by less three dirhems is nine dirhems. You (41) have, therefore, four-ninths of a square and nine dirhems less four roots, which are equal to one root. Add the four roots to the one root, then you have five roots, which are equal to four-ninths of a square and nine dirhems. Complete now your square; that is, multiply the four-ninths of a square by two and a fourth, which gives one square; multiply likewise the nine dirhems by two and a quarter; this gives twenty and a quarter; finally, multiply the five roots by two and a quarter; this gives eleven roots and a quarter. You have, therefore, a square and twenty dirhems and a quarter, equal to eleven roots and a quarter. Reduce this according to what I taught you about halving the roots.

---

\* $[x-(\frac{x}{3}+3)]^2 = x$
or $[\frac{2x}{3}-3]^2 = x$
$\frac{4x^2}{9} + 9 = 5x$
$x^2 + 20\frac{1}{4} = 11\frac{1}{4}x$
$x = 9$, or $2\frac{1}{4}$

( 58 )

If the instance be: "To find a number,* one-third of which, when multiplied by one-fourth of it, restores the *number,"† then the computation is: You multiply one-third of thing by one-fourth of thing, this gives one-twelfth of a square, equal to thing, and the square is equal to twelve things, which is the root of one hundred and forty-four.

If the instance be: "A number,* one-third of which and one dirhem multiplied by one-fourth of it and two dirhems restore the number,* with a surplus of thirteen dirhems;"‡ then the computation is this: You multiply one-third of thing by one-fourth of thing, this gives half one-sixth of a square; and you multiply two dirhems by one-third of thing, this gives two-thirds of a root; and one dirhem by one-fourth of thing gives one-fourth of a root; and one dirhem by two dirhems gives two dirhems. This altogether is one-twelfth of a square and two dirhems and eleven-(42) twelfths of a thing, equal to thing and thirteen dir-

---

* "Square" in the original.

† $\frac{x}{3} \times \frac{x}{4} = x$
$x^2 = 12x$
$x = 12$

‡ $\left(\frac{x}{3}+1\right)\cdot\left(\frac{x}{4}+2\right) = x+13$
$\frac{x^2}{12} + \frac{11}{12}x + 2 = x + 13$
$\frac{x^2}{12} = \frac{x}{12} + 11$
$x^2 = x + 132$
$x = \frac{1}{2} + \frac{23}{2} = 12$

hems. Remove now two dirhems from thirteen, on account of the other two dirhems, the remainder is eleven dirhems. Remove then the eleven-twelfths of a root from the one (root on the opposite side), there remains one-twelfth of a root and eleven dirhems, equal to one-twelfth of a square. Complete the square: that is, multiply it by twelve, and do the same with all you have. The product is a square, which is equal to a hundred and thirty-two dirhems and one root. Reduce this, according to what I have taught you, it will be right.

If the instance be: "A dirhem and a half to be divided among one person and certain persons, so that the share of the one person be twice as many dirhems as there are other persons;"* then the Computation is this:† You say, the one person and some persons are one and thing: it is the same as if the question had been one dirhem and a half to be divided by one and thing, and the share of one person to be equal to two things. Multiply, therefore, two things by one and

---

* The enunciation in the original is faulty, and I have altered it to correspond with the computation. But in the computation, $x$, the number of persons, is fractional! I am unable to correct the passage satisfactorily.

$$† \frac{1\frac{1}{2}}{1+x} = 2x$$
$$x2 + x = \tfrac{3}{4}$$
$$x = 1 - \tfrac{1}{2}$$
$$x = \tfrac{1}{2}$$

thing; it is two squares and two things, equal to one dirhem and a half. Reduce them to one square: that is, take the moiety of all you have. You say, therefore, one square and one thing are equal to three-fourths of a dirhem. Reduce this, according to what I have taught you in the beginning of this work.

If the instance be: "A number,* you remove one-third of it, and one-fourth of it, and four dirhems: then you multiply the remainder by itself, and the number,* is restored, with a surplus of twelve dirhems:' then the computation is this: You take thing, and subtract from it one-third and one-fourth; there remain five-twelfths of thing. Subtract from this four dirhems: the remainder is five-twelfths of thing less four dirhems. Multiply this by itself. Thus the five parts become five-and-twenty parts; and if you multiply twelve by itself, it is a hundred and forty four. This makes, therefore, five and twenty hundred and forty-fourths of a square. Multiply then the four dirhems twice by the five-twelfths; this gives forty parts, every twelve of which make one root (forty-twelfths); finally, the four

---

* "Square" in the original.

† $(x - \frac{1}{3}x - \frac{1}{4}x - 4)^2 = x + 12$

$(\frac{5}{12}x - 4)^2 = x + 12$

$\frac{25}{144}x^2 + 16 - 3\frac{1}{3}x = x + 12$

$\frac{25}{144}x^2 + 4 = 4\frac{1}{3}x$

$x^2 + 23\frac{1}{25} = 24\frac{24}{25}x$

$\sqrt{\left[\left(\frac{24\frac{24}{25}}{2}\right)^2 - 23\frac{1}{25}\right]} + \frac{24\frac{24}{25}}{2} - x$

$11\frac{13}{25} + 12\frac{12}{25} = 24 = x$

dirhems, multiplied by four dirhems, give sixteen dirhems to be added. The forty-twelfths are equal to three roots and one-third of a root, to be subtracted. The whole product is, therefore, twenty-five-hundred-and-forty-fourths of a square and sixteen dirhems less three roots and one-third of a root, equal to the original number,* which is thing and twelve dirhems. Reduce this, by adding the three roots and one-third to the thing and twelve dirhems. Thus you have four roots and one-third of a root and twelve dirhems. Go on balancing, and subtract the twelve (dirhems) from sixteen; there remain four dirhems and five-and-twenty-hundred-and-forty-fourths of a square, equal to four roots and one-third. Now it is necessary to complete the square. This you can accomplish by multiplying all you have by five and nineteen twenty-fifths. Multiply, therefore, the twenty-five-one-hundred-and-forty-fourths of a square by five and nineteen twenty-fifths. This gives a square. Then multiply the four (44) dirhems by five and nineteen twenty-fifths; this gives twenty-three dirhems and one twenty-fifth. Then multiply four roots and one-third by five and nineteen twenty-fifths; this gives twenty-four roots and twenty-four twenty-fifths of a root. Now halve the number of the roots: the moiety is twelve roots and twelve twenty-fifths of a root. Multiply this by itself. It is one hundred-and-fifty-five dirhems and four hundred-and-

---

\* " Square " in the original.

sixty-nine six-hundred-and-twenty-fifths. Subtract from this the twenty-three dirhems and the one twenty-fifth connected with the square. The remainder is one-hundred-and-thirty-two and four-hundred-and-forty six-hundred-and-twenty-fifths. Take the root of this: it is eleven dirhems and thirteen twenty-fifths. Add this to the moiety of the roots, which was twelve dirhems and twelve twenty-fifths. The sum is twenty-four. It is the number* which you sought. When you subtract its third and its fourth and four dirhems, and multiply the remainder by itself, the number * is restored, with a surplus of twelve dirhems.

(45) If the question be: " To find a square-root,* which, when multiplied by two-thirds of itself, amounts to five;"† then the computation is this: You multiply one thing by two-thirds of thing; the product is two-thirds of square, equal to five. Complete it by adding its moiety to it, and add to five likewise its moiety. Thus you have a square, equal to seven and a half. Take its root; it is the thing which you required, and which, when multiplied by two-thirds of itself, is equal to five.

If the instance be: " Two numbers,‡ the difference

---

\* " Square" in the original.

† $x \times \frac{2}{3}x = 5$
$\frac{2}{3}x^2 = 5$
$x^2 = 7\frac{1}{2}$
$x = \sqrt{7\frac{1}{2}}$

‡ " Squares" in the original.

of which is two dirhems; you divide the small one by the great one, and the quotient is equal to half a dirhem;* then the computation is this: Multiply thing and two dirhems by the quotient, that is a half. The product is half a thing and one dirhem, equal to thing. Remove now half a dirhem on account of the half dirhem on the other side. The remainder is one dirhem, equal to half a thing. Double it: then you have thing, equal to two dirhems. This is one of the two numbers,† and the other is four.

Instance: "You divide one dirhem amongst a certain number of men, which number is thing. Now you add one man more to them, and divide again one dirhem amongst them; the quota of each is then one-sixth of a dirhem less than at the first time."‡ Computation: You multiply the first number of men, which is thing, by the difference of the share for each of the other number. Then multiply the product by the first and second number of men, and divide the product by the

---

\* $\frac{x}{x+2} = \frac{1}{2}$
$\frac{1}{2}x + 1 = x$
$\frac{1}{2}x = 1$
$x = 2, \ x + 2 = 4$

† "Squares" in the original.

‡ $\frac{1}{x} - \frac{1}{x+1} = \frac{1}{6}$
$1 = \frac{x(x+1)}{6}$
$x^2 + x = 6$
$\sqrt{[\frac{1}{2}]^2 + 6} - \frac{1}{2} = x = 2$

difference of these two numbers. Thus you obtain the sum which shall be divided. Multiply, therefore, the first number of men, which is thing, by the one-sixth, which is the difference of the shares; this gives one-sixth of root. Then multiply this by the original number of the men, and that of the additional one, that is to say, by thing plus one. The product is one-sixth of square and one-sixth of root divided by one (46) dirhem, and this is equal to one dirhem. Complete the square which you have through multiplying it by six. Then you have a square and a root equal to six dirhems. Halve the root and multiply the moiety by itself, it is one-fourth. Add this to the six; take the root of the sum and subtract from it the moiety of the root, which you have multiplied by itself, namely, a half. The remainder is the first number of men; which in this instance is two.

If the instance be: "To find a square-root,* which when multiplied by two-thirds of itself amounts to five:"† then the computation is this: If you multiply it by itself, it gives seven and a half. Say, therefore,

---

\* " Square " in the original.

† $\frac{2}{3}x^2 = 5$
$x^2 = 7\frac{1}{2}$
$x = \sqrt{7\frac{1}{2}}$
$\sqrt{7\frac{1}{2}} \times \frac{2}{3}\sqrt{7\frac{1}{2}} = 5$
$\sqrt{\frac{4}{9} \times 7\frac{1}{2}} = \sqrt{3\frac{1}{3}} = \frac{2}{3}\sqrt{7\frac{1}{2}}$
$\sqrt{3\frac{1}{3} \times 7\frac{1}{2}} = \sqrt{25} = 5$

it is the root of seven and a half multiplied by two-thirds of the root of seven and a half. Multiply then two-thirds by two-thirds, it is four-ninths; and four-ninths multiplied by seven and a half is three and a third. The root of three and a third is two-thirds of the root of seven and a half. Multiply three and a third by seven and a half; the product is twenty-five, and its root is five.

If the instance be: "A square multiplied by three of its roots is equal to five times the original square;"* then this is the same as if it had been said, a square, which when multiplied by its root, is equal to the first square and two-thirds of it. Then the root of the square is one and two-thirds, and the square is two dirhems and seven-ninths.

If the instance be: "Remove one-third from a square, then multiply the remainder by three roots of the first square, and the first square will be restored."†
Computation: If you multiply the first square, before removing two-thirds from it, by three roots of the same, then it is one square and a half; for according to the statement two-thirds of it multiplied by three

---

\* $x^2 \times 3x = 5x^2$
$x^2 \times x = 1\frac{2}{3}x^2$
$x = 1\frac{2}{3}$
$x^2 = 2\frac{7}{9}$

† $(x^2 - \frac{1}{3}x^2) \times 3x = x^2 \therefore \frac{2}{3}x^2 \times 3x = x^2$
$x^2 \times 3x = 1\frac{1}{2}x^2$
$x = \frac{1}{2} \therefore x^2 = \frac{1}{4}$

roots give one square; and, consequently, the whole of it multiplied by three roots of it gives one square and a half. This entire square, when multiplied by one root, gives half a square; the root of the square must therefore be a half, the square one-fourth, two-thirds of the square one-sixth, and three roots of the square one and a half. If you multiply one-sixth by one and a half, the product is one-fourth, which is the square.

Instance: "A square; you subtract four roots of the same, then take one-third of the remainder; this is equal to the four roots." The square is two hundred and fifty-six.* Computation: You know that one-third of the remainder is equal to four roots; consequently, the whole remainder must be twelve roots; add to this the four roots; the sum is sixteen, which is the root of the square.

Instance: "A square; you remove one root from it; and if you add to this root a root of the remainder, the sum is two dirhems."† Then, this is the root of a

---

\* $\frac{x^2 - 4x}{3} = 4x$

$x^2 - 4x = 12x$

$x^2 = 16x$

$x = 16 \therefore x^2 = 256$

† $\sqrt{x^2 - x} + x = 2$

$\sqrt{x^2 - x} = 2 - x$

$x^2 - x = 4 + x^2 - 4x$

$x^2 + 3x = 4 + x^2$

$3x = 4$

$x = 1\frac{1}{3}$

square, which, when added to the root of the same square, less one root, is equal to two dirhems. Subtract from this one root of the square, and subtract also from the two dirhems one root of the square. Then two dirhems less one root multiplied by itself is four dirhems and one square less four roots, and this is equal to a square less one root. Reduce it, and you find a square and four dirhems, equal to a square and three roots. Remove square by square; there remain three roots, equal to four dirhems; consequently, one root is equal to one dirhem and one-third. This is the root of the square, and the square is one dirhem and seven-ninths of a dirhem. (48)

Instance : " Subtract three roots from a square, then multiply the remainder by itself, and the square is restored."* You know by this statement that the remainder must be a root likewise; and that the square consists of four such roots; consequently, it must be sixteen.

---

$$* \; (x^2 - 3x)^2 = x^2$$
$$x^2 - 3x = x$$
$$x^2 = 4x$$
$$x = 4$$

( 68 )

## ON MERCANTILE TRANSACTIONS.

You know that all mercantile transactions of people, such as buying and selling, exchange and hire, comprehend always two notions and four numbers, which are stated by the enquirer; namely, measure and price, and quantity and sum. The number which expresses the measure is inversely proportionate to the number which expresses the sum, and the number of the price inversely proportionate to that of the quantity. Three of these four numbers are always known, one is unknown, and this is implied when the person inquiring says *how much?* and it is the object of the question. The computation in such instances is this, that you try the three given numbers; two of them must necessarily be inversely proportionate the one to the other. Then you multiply these two proportionate numbers by each other, and you divide the product by the third given number, the proportionate of which is unknown. The quotient of this division is the unknown number, which the inquirer asked for; and it is inversely proportionate to the divisor.*

*Examples.—For the first case:* If you are told, " ten (49) for six, how much for four?" then *ten* is the measure;

---

\* If $a$ is given for $b$, and $A$ for $B$, then $a : b :: A : B$ or $aB = bA \therefore a = \frac{bA}{B}$ and $b = \frac{aB}{A}$.

*six* is the price; the expression *how much* implies the unknown number of the quantity; and *four* is the number of the sum. The number of the measure, which is *ten*, is inversely proportionate to the number of the sum, namely, *four*. Multiply, therefore, ten by four, that is to say, the two known proportionate numbers by each other; the product is forty. Divide this by the other known number, which is that of the price, namely, six. The quotient is six and two-thirds; it is the unknown number, implied in the words of the question "*how much?*" it is the quantity, and inversely proportionate to the six, which is the price.

*For the second case:* Suppose that some one ask this question: "ten for eight, what must be the sum for four?" This is also sometimes expressed thus: "What must be the price of four of them?" Ten is the number of the measure, and is inversely proportionate to the unknown number of the sum, which is involved in the expression *how much* of the statement. Eight is the number of the price, and this is inversely proportionate to the known number of the quantity, namely, four. Multiply now the two known proportionate numbers one by the other, that is to say, four by eight. The product is thirty-two. Divide this by the other known number, which is that of the measure, namely, ten. The quotient is three and one-fifth; this is the number of the sum, and inversely proportionate to the ten which was the divisor. In this manner all computations in matters of business may be solved.

If somebody says, "a workman receives a pay of ten (50) dirhems per month ; how much must be his pay for six days?" Then you know that six days are one-fifth of the month; and that his portion of the dirhems must be proportionate to the portion of the month. You calculate it by observing that one month, or thirty days, is the measure, ten dirhems the price, six days the quantity, and his portion the sum. Multiply the price, that is, ten, by the quantity, which is proportionate to it, namely, six; the product is sixty. Divide this by thirty, which is the known number of the measure. The quotient is two dirhems, and this is the sum.

This is the proceeding by which all transactions concerning exchange or measures or weights are settled.

## MENSURATION.

Know that the meaning of the expression "one by one" is mensuration : one yard (in length) by one yard (in breadth) being understood.

Every quadrangle of equal sides and angles, which has one yard for every side, has also *one* for its area. Has such a quadrangle two yards for its side, then the area of the quadrangle is four times the area of a quadrangle, the side of which is one yard. The same takes place with three by three, and so on, ascending or descending : for instance, a half by a half, which gives a quarter, or other fractions, always following the same rule. A quadrate, every side of which is half a yard, is equal to one-fourth of the figure which has one yard for its side. In the same manner, one-third by one-third, or one-fourth by one-fourth, or one-fifth by one-fifth, or two-thirds by a half, or more or less than this, always according to the same rule. (51)

One side of an equilateral quadrangular figure, taken once, is its root; or if the same be multiplied by two, then it is like two of its roots, whether it be small or great.

If you multiply the height of any equilateral triangle by the moiety of the basis upon which the line marking the height stands perpendicularly, the product gives the area of that triangle.

In every equilateral quadrangle, the product of one diameter multiplied by the moiety of the other will be equal to the area of it.

In any circle, the product of its diameter, multiplied by three and one-seventh, will be equal to the periphery. This is the rule generally followed in practical life, though it is not quite exact. The geometricians have two other methods. One of them is, that you multiply the diameter by itself; then by ten, and hereafter take the root of the product; the root will be the periphery. The other method is used by the astronomers among them: it is this, that you multiply the diameter by sixty-two thousand eight hundred and thirty-two and then divide the product by twenty thousand; the quotient is the periphery. Both methods come very nearly to the same effect.*

If you divide the periphery by three and one-seventh, the quotient is the diameter.

(52) The area of any circle will be found by multiplying the moiety of the circumference by the moiety of the diameter; since, in every polygon of equal sides and angles, such as triangles, quadrangles, pentagons, and so on, the area is found by multiplying the moiety of the circumference by the moiety of the diameter of the middle circle that may be drawn through it.

If you multiply the diameter of any circle by itself, and subtract from the product one-seventh and half one-seventh of the same, then the remainder is equal to the area of the circle. This comes very nearly to the same result with the method given above. †

Every part of a circle may be compared to a bow. It must be either exactly equal to half the circumference, or less or greater than it. This may be ascertained by the arrow of the bow. When this becomes equal to the moiety of the chord, then the arc is

---

* The three formulas are,

    1st, $3\tfrac{1}{7}d = p$  i.e. $3.1428\,d$

    2d, $\sqrt{10d^2} = p$  i.e. $3.16227\,d$

    3d, $\dfrac{d \times 62832}{20000} = p$  i.e. $3.1416\,d$

† The area of a circle whose diameter is $d$ is $\pi\dfrac{d^2}{4} = \dfrac{22}{7 \times 4}d^2 = \left(1 - \tfrac{1}{7} - \tfrac{1}{2 \times 7}\right)d^2.$

exactly the moiety of the circumference: is it shorter than the moiety of the chord, then the bow is less than half the circumference; is the arrow longer than half the chord, then the bow comprises more than half the circumference.

If you want to ascertain the circle to which it belongs, multiply the moiety of the chord by itself, divide it by the arrow, and add the quotient to the arrow, the sum is the diameter of the circle to which this bow belongs.

If you want to compute the area of the bow, multiply the moiety of the diameter of the circle by the moiety of the bow, and keep the product in mind. Then subtract the arrow of the bow from the moiety of the diameter of the circle, if the bow is smaller than half the circle; or if it is greater than half the circle, subtract half the diameter of the circle from the arrow of the bow. Multiply the remainder by the moiety of the chord of the bow, and subtract the product from that which you have kept in mind if the bow is smaller (53) than the moiety of the circle, or add it thereto if the bow is greater than half the circle. The sum after the addition, or the remainder after the subtraction, is the area of the bow.

The bulk of a quadrangular body will be found by multiplying the length by the breadth, and then by the height.

If it is of another shape than the quadrangular (for instance, circular or triangular), so, however, that a line representing its height may stand perpendicularly on its basis, and yet be parallel to the sides, you must calculate it by ascertaining at first the area of its basis. This, multiplied by the height, gives the bulk of the body.

Cones and pyramids, such as triangular or quadrangular ones, are computed by multiplying one-third of the area of the basis by the height.

Observe, that in every rectangular triangle the two short sides, each multiplied by itself and the products added together, equal the product of the long side multiplied by itself.

The proof of this is the following. We draw a quadrangle, with equal sides and angles A B C D. We divide the line A C into two moieties in the point H, from which we draw a parallel to the point R. Then we divide, also, the line A B into two moieties at the point T, and draw a parallel to the point G. Then the quadrate A B C D is divided into four quadrangles of equal sides and angles, and of equal area; namely, the squares A K, C K, B K, and D K. Now, we draw from (54) the point H to the point T a line which divides the quadrangle A K into two equal parts: thus there arise two triangles from the quadrangle, namely, the triangles A T H and H K T. We know that A T is the moiety of A B, and that A H is equal to it, being the moiety of A C; and the line T H joins them opposite the right angle. In the same manner we draw lines from T to R, and from R to G, and from G to H. Thus from all the squares eight equal triangles arise, four of which must, consequently, be equal to the moiety of the great quadrate A D. We know that the line A T multiplied by itself is like the area of two triangles, and A K gives the area of two triangles equal to them; the sum of them is therefore four triangles. But the line HT, multiplied by itself, gives likewise the area of four such triangles. We perceive, therefore, that the sum of AT multiplied by itself, added to AH multiplied by itself, is equal to TH multiplied by itself. This is the observation which we were desirous to elucidate. Here is the figure to it:

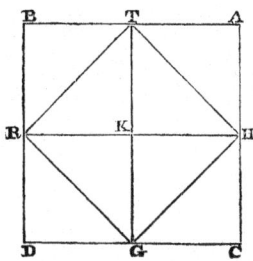

Quadrangles are of five kinds: firstly, with right (55) angles and equal sides; secondly, with right angles and unequal sides; thirdly, the rhombus, with equal sides and unequal angles; fourthly, the rhomboid, the length of which differs from its breadth, and the angles of which are unequal, only that the two long and the two short sides are respectively of equal length; fifthly, quadrangles with unequal sides and angles.

*First kind.*—The area of any quadrangle with equal sides and right angles, or with unequal sides and right angles, may be found by multiplying the length by the breadth. The product is the area. For instance: a quadrangular piece of ground, every side of which has five yards, has an area of five-and-twenty square yards. Here is its figure.

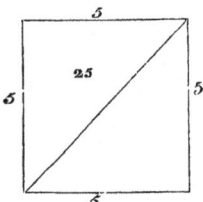

*Second kind.*—A quadrangular piece of ground, the two long sides of which are of eight yards each, while the breadth is six. You find the area by multiplying six by eight, which yields forty-eight yards. Here is (56) the figure to it:

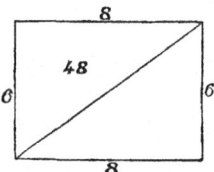

*Third kind,* the Rhombus.—Its sides are equal: let each of them be five, and let its diagonals be, the one eight and the other six yards. You may then compute the area, either from one of the diagonals, or from both. As you know them both, you multiply the one by the moiety of the other, the product is the area: that is to say, you multiply eight by three, or six by four; this yields twenty-four yards, which is the area.

If you know only one of the diagonals, then you are aware, that there are two triangles, two sides of each of which have every one five yards, while the third is the diagonal. Hereafter you can make the computation according to the rules for the triangles.* This is the figure:

*The fourth kind*, or Rhomboid, is computed in the same way as the rhombus. Here is the figure to it:

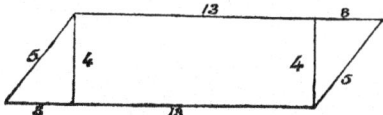

The other quadrangles are calculated by drawing a (57) diagonal, and computing them as triangles.

Triangles are of three kinds, acute-angular, obtuse-angular, or rectangular. The peculiarity of the rectangular triangle is, that if you multiply each of its two short sides by itself, and then add them together, their sum will be equal to the long side multiplied by itself. The character of the acute-angled triangle is

---

* If the two diagonals are $d$ and $d'$, and the side $s$, the area of the rhombus is $\frac{dd'}{2} = d \times \sqrt{s^2 - \frac{d^2}{4}}$.

this: if you multiply every one of its two short sides by itself, and add the products, their sum is more than the long side alone multiplied by itself. The definition of the obtuse-angled triangle is this: if you multiply its two short sides each by itself, and then add the products, their sum is less than the product of the long side multiplied by itself.

The rectangular triangle has two cathetes and an hypotenuse. It may be considered as the moiety of a quadrangle. You find its area by multiplying one of its cathetes by the moiety of the other. The product is the area.

*Examples.*—A rectangular triangle; one cathete being (58) six yards, the other eight, and the hypotenuse ten. You make the computation by multiplying six by four: this gives twenty-four, which is the area. Or if you prefer, you may also calculate it by the height, which rises perpendicularly from the longest side of it: for the two short sides may themselves be considered as two heights. If you prefer this, you multiply the height by the moiety of the basis. The product is the area. This is the figure:

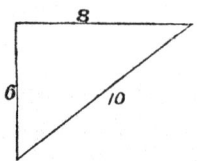

*Second kind.*—An equilateral triangle with acute angles, every side of which is ten yards long. Its area may be ascertained by the line representing its height and the point from which it rises.* Observe, that in every isosceles triangle, a line to represent the height drawn to the basis rises from the latter in a right angle, and the point from which it proceeds is always situated in the midst of the basis; if, on the contrary, the two sides are not equal, then this point never lies in the middle of the basis. In the case now before us we perceive, that towards whatever side we may draw the line which is to represent the height, it must necessarily always fall in the middle of it, where the length of the basis is five. Now the height will be ascertained thus. You multiply five by itself; then multiply one of the sides, that is ten, by itself, which gives a hundred. Now you subtract from this the product of five multiplied by itself, which is twenty-five. (59) The remainder is seventy-five, the root of which is the height. This is a line common to two rectangular triangles. If you want to find the area, multiply the root of seventy-five by the moiety of the basis, which is five. This you perform by multiplying at first five by itself; then you may say, that the root of seventy-five is to be multiplied by the root of twenty-five. Multiply seventy-five by twenty-five. The product is one thousand eight hundred and seventy-five; take its root, it is

---

\* The height of the equilateral triangle whose side is 10, is $\sqrt{10^2 - 5^2} = \sqrt{75}$, and the area of the triangle is $5\sqrt{75} = 25\sqrt{3}$

( 80 )

the area: it is forty-three and a little.*  Here is the figure:

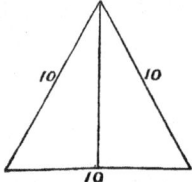

There are also acute-angled triangles, with different sides. Their area will be found by means of the line representing the height and the point from which it proceeds. Take, for instance, a triangle, one side of which is fifteen yards, another fourteen, and the third thirteen yards. In order to find the point from which the line marking the height does arise, you may take for the basis any side you choose; e. g. that which is fourteen yards long. The point from which the line (60) representing the height does arise, lies in this basis at an unknown distance from either of the two other sides. Let us try to find its unknown distance from the side which is thirteen yards long. Multiply this distance by itself; it becomes an [unknown] square. Subtract this from thirteen multiplied by itself; that is, one hundred and sixty-nine. The remainder is one hundred and sixty-nine less a square. The root from this is the height. The remainder of the basis is fourteen less thing. We multiply this by itself; it becomes one hundred and ninety-six, and a square less twenty-

---

* The root is 43. 3 +

eight things. We subtract this from fifteen multiplied by itself; the remainder is twenty-nine dirhems and twenty-eight things less one square. The root of this is the height. As, therefore, the root of this is the height, and the root of one hundred and sixty-nine less square is the height likewise, we know that they both are the same.* Reduce them, by removing square against square, since both are negatives. There remain twenty-nine [dirhems] plus twenty-eight things, which are equal to one hundred and sixty-nine. Subtract now twenty-nine from one hundred and sixty-nine. The remainder is one hundred and forty, equal to twenty-eight things. One thing is, consequently, five. This is the distance of the said point from the side of thirteen yards. The complement of the basis towards the other side is nine. Now in order to find the height, you multiply five by itself, and subtract it from the contiguous side, which is thirteen, multiplied by itself. The remainder is one hundred and forty-four. Its root is the height. It is twelve. The height forms always two right angles with the basis, and it is called the *column*, on account of its standing perpendicularly. Multiply the height into half the basis, which is seven. The

---

$$* \sqrt{169 - x^2} = 29 + 28x - x^2$$
$$163 = 29 + 28x$$
$$140 = 28x$$
$$5 = x$$

product is eighty-four, which is the area. Here is the figure :

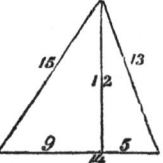

*The third species* is that of the obtuse-angled triangle with one obtuse angle and sides of different length. For instance, one side being six, another five, and the third nine. The area of such a triangle will be found by means of the height and of the point from which a line representing the same arises. This point can, within such a triangle, lie only in its longest side. Take therefore this as the basis : for if you choose to take one of the short sides as the basis, then this point would fall beyond the triangle. You may find the distance of this point, and the height, in the same manner, which I have shown in the acute-angled triangle; the whole computation is the same. Here is the figure:

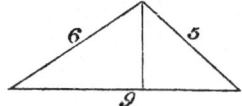

(62) We have above treated at length of the circles, of their qualities and their computation. The following is an example: If a circle has seven for its diameter, then it has twenty-two for its circumference. Its area you find in the following manner : Multiply the moiety

of the diameter, which is three and a half, by the moiety of the circumference, which is eleven. The product is thirty-eight and a half, which is the area. Or you may also multiply the diameter, which is seven, by itself: this is forty-nine; subtracting herefrom one-seventh and half one-seventh, which is ten and a half, there remain thirty-eight and a half, which is the area. Here is the figure:

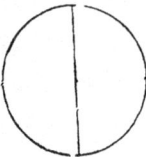

If some one inquires about the bulk of a pyramidal pillar, its base being four yards by four yards, its height ten yards, and the dimensions at its upper extremity two yards by two yards; then we know already that every pyramid is decreasing towards its top, and that one-third of the area of its basis, multiplied by the height, gives its bulk. The present pyramid has no top. We must therefore seek to ascertain what is wanting in its height to complete the top. We observe, that the proportion of the entire height to the ten, which we have now before us, is equal to the proportion of four to two. Now as two is the moiety of four, ten must likewise be the moiety of the entire height, and the whole height of the pillar must be twenty yards. At present we take one-third of the area of the basis, that is, five and one-third, and multiply it by the length, which is twenty. The product is one hundred (63)

and six yards and two-thirds. Herefrom we must then subtract the piece, which we have added in order to complete the pyramid. This we perform by multiplying one and one-third, which is one-third of the product of two by two, by ten: this gives thirteen and a third. This is the piece which we have added in order to complete the pyramid. Subtracting this from one hundred and six yards and two-thirds, there remain ninety-three yards and one-third: and this is the bulk of the mutilated pyramid. This is the figure:

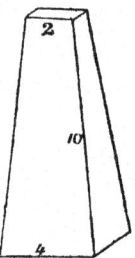

If the pillar has a circular basis, subtract one-seventh and half a seventh from the product of the diameter multiplied by itself, the remainder is the basis.

If some one says: "There is a triangular piece of land, two of its sides having ten yards each, and the basis twelve; what must be the length of one side of a quadrate situated within such a triangle?" the solution is this. At first you ascertain the height of the triangle, by multiplying the moiety of the basis, (which is six) by itself, and subtracting the product, which is thirty-six, from one of the two short sides multiplied by itself, which is one hundred; the remainder is sixty-four: take the root from this; it is eight. This (64) is the height of the triangle. Its area is, therefore, forty-eight yards: such being the product of the height multiplied by the moiety of the basis, which is six. Now we assume that one side of the quadrate inquired for is thing. We multiply it by itself; thus it becomes a square, which we keep in mind. We know that there must remain two triangles on the two sides of the quadrate, and one above it. The two triangles on both sides of it are equal to each other: both having the same height and being rectangular. You find their area by multiplying thing by six less half a thing, which gives six things less half a square. This is the area of both the triangles on the two sides of the quadrate together. The area of the upper triangle will be found by multiplying eight less thing, which is the height, by half one thing. The product is four things less half a square. This altogether is equal to the area of the quadrate plus that of the three triangles: or, ten things are equal to forty-eight, which is the area of the great triangle. One thing from this is four yards and four-fifths of a yard; and this is the length of any side of the quadrate. Here is the figure:

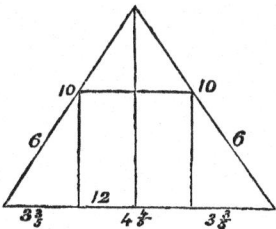

## ON LEGACIES.

### *On Capital, and Money lent.*

(65) "A MAN dies, leaving two sons behind him, and bequeathing one-third of his capital to a stranger. He leaves ten dirhems of property and a claim of ten dirhems upon one of the sons."

Computation: You call the sum which is taken out of the debt thing. Add this to the capital, which is ten dirhems. The sum is ten and thing. Subtract one-third of this, since he has bequeathed one-third of his property, that is, three dirhems and one-third of thing. The remainder is six dirhems and two-thirds of thing. Divide this between the two sons. The portion of each of them is three dirhems and one-third plus one-third of thing. This is equal to the thing which was sought for.* Reduce it, by removing one-third from

---

* If a father dies, leaving $n$ sons, one of whom owes the father a sum exceeding an $n$th part of the residue of the father's estate, after paying legacies, then such son retains the whole sum which he owes the father: part, as a set-off against his share of the residue, the surplus as a gift from the father.

In the present example, let each son's share of the residue be equal to $x$.

$\frac{2}{3}[10+x] = 2x$  ∴  $1 + x - 3x$  ∴  $10 = 2x$  ∴  $x = 5$.
The stranger receives 5; and the son, who is not indebted to the father, receives 5.

thing, on account of the other third of thing. There remain two-thirds of thing, equal to three dirhems and one-third. It is then only required that you complete the thing, by adding to it as much as one half of the same; accordingly, you add to three and one-third as much as one-half of them: This gives five dirhems, which is the thing that is taken out of the debts.

If he leaves two sons and ten dirhems of capital and a demand of ten dirhems against one of the sons, and bequeaths one-fifth of his property and one dirhem to a stranger, the computation is this: Call the sum which is taken out of the debt, thing. Add this to the property; the sum is thing and ten dirhems. Subtract one-fifth of this, since he has bequeathed one-fifth of (66) his capital, that is, two dirhems and one-fifth of thing; the remainder is eight dirhems and four-fifths of thing. Subtract also the one dirhem which he has bequeathed; there remain seven dirhems and four-fifths of thing. Divide this between the two sons; there will be for each of them three dirhems and a half plus two-fifths of thing; and this is equal to one thing.* Reduce it by subtracting two-fifths of thing from thing. Then you have three-fifths of thing, equal to three dirhems and a half. Complete the thing by adding to it two-thirds of the same: add as much to the three dirhems and a half,

---

* $\frac{4}{5}[10+x]-1=2x$  ∴  $\frac{2}{5}[10+x]-\frac{1}{2}=x$
∴ $3\frac{1}{2}=\frac{3}{5}x$  ∴  $x=\frac{35}{6}=5\frac{5}{6}$
The stranger receives $\frac{1}{5}[10+\frac{35}{6}]+1=4\frac{1}{6}$

namely, two dirhems and one-third; the sum is five and five-sixths. This is the thing, or the amount which is taken from the debt.

If he leaves three sons, and bequeaths one-fifth of his property less one dirhem, leaving ten dirhems of capital and a demand of ten dirhems against one of the sons, the computation is this: You call the sum which is taken from the debt thing. Add this to the capital; it gives ten and thing. Subtract from this one-fifth of it for the legacy: it is two dirhems and one-fifth of thing. There remain eight dirhems and four-fifths of thing; add to this one dirhem, since he stated "less one dirhem." Thus you have nine dirhems and four-fifths of thing. Divide this between the three sons. There will be for each son three dirhems, and one-fifth and one-third and one-fifth of thing. This equals one thing.* Subtract one-fifth and one-third of one-fifth of thing from thing. There remain eleven-fifteenths of thing, equal to three dirhems. It is now required to complete the thing. For this purpose, add to it four-elevenths, and do the same with the three dirhems, by adding to them one dirhem and one-eleventh. Then you have four dirhems and one-eleventh, which are equal to thing. This is the sum which is taken out of the debt.

---

* $\frac{4}{5}[10+x]+1=3x$ ∴ $9=2\frac{1}{5}x$ ∴ $\frac{45}{11}$ or $4\frac{1}{11}=x$
The stranger receives $\frac{1}{5}[10+\frac{45}{11}]-1=1\frac{9}{11}$

## On another Species of Legacy.

" A man dies, leaving his mother, his wife, and two brothers and two sisters by the same father and mother with himself; and he bequeaths to a stranger one-ninth of his capital."

Computation:* You constitute their shares by taking them out of forty-eight parts. You know that if you take one-ninth from any capital, eight-ninths of it will remain. Add now to the eight-ninths one-eighth of the same, and to the forty-eight also one-eighth of them, namely, six, in order to complete your capital. This gives fifty-four. The person to whom one-ninth is bequeathed receives six out of this, being one-ninth of the whole capital. The remaining forty-eight will be distributed among the heirs, proportionably to their legal shares.

If the instance be: "A woman dies, leaving her husband, a son, and three daughters, and bequeathing

---

* It appears in the sequel (p. 96) that a widow is entitled to $\frac{1}{8}$th, and a mother to $\frac{1}{6}$th of the residue; $\frac{1}{8}+\frac{1}{6}=\frac{14}{48}$, leaving $\frac{34}{48}$ of the residue to be distributed between two brothers and two sisters; that is, $\frac{17}{48}$ between a brother and a sister; but in what proportion these 17 parts are to be divided between the brother and sister does not appear in the course of this treatise.

Let the whole capital of the testator $= 1$
and let each 48th share of the residue $= x$
$\frac{8}{9} = 48x \quad \therefore \frac{1}{9} = 6x \quad \therefore \frac{1}{54} = x$

that is, each 48th part of the residue $=\frac{1}{54}$th of the whole capital.

(68) to a stranger one-eighth and one-seventh of her capital;" then you constitute the shares of the heirs, by taking them out of twenty.* Take a capital, and subtract from it one-eighth and one-seventh of the same. The remainder is, a capital less one-eighth and one-seventh. Complete your capital by adding to that which you have already, fifteen forty-one parts. Multiply the parts of the capital, which are twenty, by forty-one; the product is eight hundred and twenty. Add to it fifteen forty-one parts of the same, which are three hundred: the sum is one thousand one hundred and twenty parts. The person to whom one-eighth and one-seventh were bequeathed, receives one-eighth and one-seventh of this. One seventh of it is one hundred and sixty, and one eighth one hundred and forty. Subtracting this, there remain eight hundred and twenty parts for the heirs, proportionably to their legal shares.

---

* A husband is entitled to $\frac{1}{4}$th of the residue, and the sons and daughters divide the remaining $\frac{3}{4}$ths of the residue in such proportion, that a son receives twice as much as a daughter. In the present instance, as there are three daughters and one son, each daughter receives $\frac{1}{5}$ of $\frac{3}{4}$, $=\frac{3}{20}$, of the residue, and the son, $\frac{6}{20}$. Since the stranger takes $\frac{1}{8}+\frac{1}{7}=\frac{15}{56}$ of the capital, the residue $=\frac{41}{56}$ of the capital, and each $\frac{1}{20}$th share of the residue $=\frac{1}{20}\times\frac{41}{56}=\frac{41}{1120}$ of the capital. The stranger, therefore, receives $\frac{15}{56}=\frac{15\times 20}{56\times 20}=\frac{300}{1120}$ of the capital.

## On another Species of Legacies,* viz.

If nothing has been imposed on some of the heirs,† and something has been imposed on others; the legacy amounting to more than one-third. It must be known, that the law for such a case is, that if more than one-third of the legacy has been imposed on one of the heirs, this enters into his share; but that also those on whom nothing has been imposed must, nevertheless, contribute one-third.

Example: " A woman dies, leaving her husband, a son, and her mother. She bequeaths to a person two-fifths, and to another one-fourth of her capital. She imposes the two legacies together on her son, and on her mother one moiety (of the mother's share of the residue); on her husband she imposes nothing but one-third, (which he must contribute, according to the

---

* The problems in this chapter may be considered as belonging rather to Law than to Algebra, as they contain little more than enunciations of the law of inheritance in certain complicated cases.

† If some heirs are, by a testator, charged with payment of bequests, and other heirs are not charged with payment of any bequests whatever: if one bequest exceeds in amount $\frac{1}{3}$d of the testator's whole property; and if one of his heirs is charged with payment of more than $\frac{1}{3}$d of such bequest; then, whatever share of the residue such heir is entitled to receive, the like share must he pay of the bequest wherewith he is charged, and those heirs whom the testator has not charged with any payment, must each contribute towards paying the bequests a third part of their several shares of the residue.

( 92 )

(69) law)."* Computation: You constitute the shares of the heritage, by taking them out of twelve parts: the son receives seven of them, the husband three, and the mother two parts. You know that the husband must give up one-third of his share; accordingly he retains twice as much as that which is detracted from his share for the legacy. As he has three parts in hand, one of these falls to the legacy, and the remaining two parts he retains for himself. The two legacies together are imposed upon the son. It is therefore necessary to subtract from his share two-fifths and one-fourth of the same. He thus retains seven twentieths of his entire original share, dividing the whole of it into twenty equal parts. The mother retains as much as she contributes to the legacy; this is one (twelfth part), the entire amount of what she had received being two parts.

---

\* If the bequests stated in the present example were charged on the heirs collectively, the husband would be entitled to $\frac{1}{4}$, the mother to $\frac{1}{6}$ of the residue : $\frac{1}{4}+\frac{1}{6}=\frac{5}{12}$; the remainder $\frac{7}{12}$ would be the son's share of the residue; but since the bequests, $\frac{2}{5}+\frac{1}{4}=\frac{13}{20}$ of the capital, are charged upon the son and mother, the law throws a portion of the charge on the husband.

The Husband contributes $\frac{1}{4} \times \frac{1}{3} = 20 \times \frac{1}{240}$, and retains $\frac{1}{4} \times \frac{2}{3} = 40 \times \frac{1}{240}$

The Mother .......... $\frac{1}{6} \times \frac{1}{2} = 20 \times \frac{1}{240}$, ....... $\frac{1}{6} \times \frac{1}{2} = 20 \times \frac{1}{240}$

The Son ........... $\frac{7}{12} \times \frac{13}{20} = 91 \times \frac{1}{240}$, ......... $\frac{7}{12} \times \frac{7}{20} = 49 \times \frac{1}{240}$

Total contributed $= \frac{131}{240}$    Total retained $= \frac{109}{240}$

$\frac{2}{5}+\frac{1}{4}=\frac{8}{20}+\frac{5}{20}=\frac{13}{20}$

The Legatee, to whom the $\frac{2}{5}$ are bequeathed, receives $\frac{8}{13} \times \frac{131}{240} = \frac{8 \times 131}{3120}$

The Legatee, to whom $\frac{1}{4}$ is bequeathed, receives $\frac{5}{13} \times \frac{131}{240} = \frac{5 \times 131}{3120}$

Take now a sum, one-fourth of which may be divided into thirds, or of one-sixth of which the moiety may be taken; this being again divisible by twenty. Such a capital is two hundred and forty. The mother receives one-sixth of this, namely, forty; twenty from this fall to the legacy, and she retains twenty for herself. The husband receives one-fourth, namely, sixty; from which twenty belong to the legacy, so that he retains forty. The remaining hundred and forty belong to the son; the legacy from this is two-fifths and one-fourth, or ninety-one; so that there remain forty-nine. The entire sum for the legacies is, therefore, one hundred and thirty-one, which must be divided among the two legatees. The one to whom two-fifths were bequeathed, receives eight-thirteenths of this; the one to whom one-fourth was devised, receives five-thirteenths. If you wish distinctly to express the shares of the two legatees, you need only to multiply (70) the parts of the heritage by thirteen, and to take them out of a capital of three thousand one hundred and twenty.

But if she had imposed on her son (payment of) the two-fifths to the person to whom the two-fifths were bequeathed, and of nothing to the other legatee; and upon her mother (payment of) the one-fourth to the person to whom one-fourth was granted, and of nothing to the other legatee; and upon her husband nothing besides the one-third (which he must according to law contribute) to both; then you know that this one-third comes to the advantage of the heirs collectively; and the legatee of the two-fifths receives eight-thirteenths, and the legatee of the one-fourth receives five-thirteenths from it. Constitute the shares as I have shown above, by taking twelve parts; the husband receives one-fourth of them, the mother one-sixth, and the son that which remains.* Computation: You know that at all events the husband must give up one-third of his share, which consists of three parts. The mother must likewise give up one-third, of which each legatee partakes according to the proportion of his legacy. Besides, she must pay to the legatee to whom one-fourth is bequeathed, and whose legacy has been imposed on her, as much as the difference between the one-fourth and his

---

\* $\frac{2}{5} + \frac{1}{4} = \frac{8+5}{20} = \frac{13}{20}$

The Husband, who would be entitled to $\frac{1}{4}$ of the residue, is not charged by the Testator with any bequest.

The Mother who would be entitled to $\frac{1}{6}$ of the residue, is charged with the payment of $\frac{1}{4}$ to the Legatee A.

The Son, who would be entitled to $\frac{7}{12}$ of the residue, is charged with payment of $\frac{2}{5}$ to the Legatee B.

The Husband contributes $\frac{1}{4} \times \frac{1}{3} = 780 \times \frac{1}{9360}$; retains $\frac{1}{4} \times \frac{2}{3} = \frac{1560}{9360}$

The Mother ... $\frac{1}{6}[\frac{1}{4} + \frac{8}{13} \times \frac{1}{3}] = 710 \times \frac{1}{9360}$; retains ...... $\frac{850}{9360}$

The Son ..... $\frac{7}{12}[\frac{2}{5} + \frac{5}{13} \times \frac{1}{3}] = 2884 \times \frac{1}{9360}$; retains ...... $\frac{2576}{9360}$

Total contributed $= \frac{4374}{9360}$; Total retained $= \frac{4986}{9360}$

The Legatee A, to whom $\frac{1}{4}$ is bequeathed, receives $\frac{5}{13} \times \frac{4374}{9360} = \frac{5 \times 4374}{964080}$

The Legatee B, to whom $\frac{2}{5}$ are bequeathed, receives $\frac{8}{13} \times \frac{4374}{9360} = \frac{8 \times 4374}{964080}$

portion of the one-third, namely, nineteen one hundred and fifty-sixths of her entire share, considering her share as consisting of one hundred and fifty-six parts. His portion of the one-third of her share is twenty parts. But what she gives him is one-fourth of her entire share, namely, thirty-nine parts. One third of her share is taken for both legacies, and besides nineteen parts which she must pay to him alone. The son gives to the legatee to whom two-fifths are bequeathed as much as the difference between two-fifths of his (the son's) share (71) and the legatee's portion of the one-third, namely, thirty-eight one hundred and ninety-fifths of his (the son's) entire share, besides the one-third of it which is taken off from both legacies. The portion which he (the legatee) receives from this one-third, is eight-thirteenths of it, namely, forty (one hundred and ninety-fifths); and what the son contributes of the two-fifths from his share is thirty-eight. These together make seventy-eight. Consequently, sixty-five will be taken from the son, as being one-third of his share, for both legacies, and besides this he gives thirty-eight to the one of them in particular. If you wish to express the parts of the heritage distinctly, you may do so with nine hundred and sixty-four thousand and eighty.

---

### On another Species of Legacies.

" A man dies, leaving four sons and his wife; and bequeathing to a person as much as the share of one

( 96 )

of the sons less the amount of the share of the widow." Divide the heritage into thirty-two parts. The widow receives one-eighth,* namely, four; and each son seven. Consequently the legatee must receive three-sevenths of the share of a son. Add, therefore, to the heritage three-sevenths of the share of a son, that is to say, three parts, which is the amount of the legacy. This gives thirty-five, from which the legatee receives three; and the remaining thirty-two are distributed among the heirs proportionably to their legal shares.

If he leaves two sons and a daughter,† and bequeaths to some one as much as would be the share of a third son, if he had one; then you must consider, what (72) would be the share of each son, in case he had three. Assume this to be seven, and for the entire heritage

---

* A widow is entitled to $\frac{1}{8}$th of the residue; therefore $\frac{7}{8}$ths of the residue are to be distributed among the sons of the testator. Let $x$ be the stranger's legacy. The widow's share $=\frac{1-x}{8}$; each son's share $=\frac{1}{4}\times\frac{7}{8}[1-x]$; and a son's share, minus the widow's share $=[\frac{7}{4}-1]\frac{1-x}{8}=\frac{3}{4}\cdot\frac{1-x}{8}$
$\therefore x=\frac{3}{4}\cdot\frac{1-x}{8}$ $\therefore x=\frac{3}{35}$; $1-x=\frac{32}{35}$ A son's share $=\frac{7}{35}$; the widow's share $=\frac{4}{35}$.

† A son is entitled to receive twice as much as a daughter. Were there three sons and one daughter, each son would receive $\frac{2}{9}$ths of the residue. Let $x$ be the stranger's legacy.
$\therefore \frac{2}{7}[1-x]=x$ $\therefore x=\frac{2}{9}$, and $1-x=\frac{7}{9}$
Each Son's share.... $=\frac{2}{9}[1-x]=\frac{2}{9}\times\frac{7}{9}=\frac{14}{45}$
The Daughter's share $=\frac{1}{9}[1-x]$ ...... $=\frac{7}{45}$
The Stranger's legacy $=\frac{2}{9}$ ............ $=\frac{10}{45}$

take a number, one-fifth of which may be divided into sevenths, and one-seventh of which may be divided into fifths. Such a number is thirty-five. Add to it two-sevenths of the same, namely, ten. This gives forty-five. Herefrom the legatee receives ten, each son fourteen, and the daughter seven.

If he leaves a mother, three sons, and a daughter, and bequeaths to some one as much as the share of one of his sons less the amount of the share of a second daughter, in case he had one; then you distribute the heritage into such a number of parts as may be divided among the actual heirs, and also among the same, if a second daughter were added to them.* Such a number is three hundred and thirty-six. The share of the second daughter, if there were one, would be thirty-five, and that of a son eighty; their difference is forty-five, and this is the legacy. Add to it three hundred and thirty-six, the sum is three hundred and eighty-one, which is the number of parts of the entire heritage.

---

* Let $x$ be the stranger's legacy; $1-x$ is the residue.
A widow's share of the residue is $\frac{1}{6}$th: there remains $\frac{5}{6}[1-x]$, to be distributed among the children.

Since there are 3 sons, and 1 daughter, a son's share is ............ } $\frac{2}{7} \times \frac{5}{6}[1-x]$

Were there 3 sons and 2 daughters, a daughter's share would be ...... } $\frac{1}{8} \times \frac{5}{6}[1-x]$

The difference $= \frac{9}{56} \times \frac{5}{6}[1-x]$

$\therefore x = \frac{45}{336}[1-x] \qquad \therefore x = \frac{45}{381}$

$1-x = \frac{336}{381}$; the widow's share $= \frac{56}{381}$

the daughter's share $= \frac{40}{381}$

( 98 )

If he leaves three sons, and bequeaths to some one as much as the share of one of his sons, less the share of a daughter, supposing he had one, plus one-third of the remainder of the one third; the computation will be this :* distribute the heritage into such a number of parts as may be divided among the actual heirs, and also among them if a daughter were added to them. Such a number is twenty-one. Were a daughter among the heirs, her share would be three, and that of a son seven. The testator has therefore bequeathed to the (73) legatee four-sevenths of the share of a son, and one-third of what remains from one-third. Take therefore one-third, and remove from it four-sevenths of the share of a son. There remains one-third of the capital less four-sevenths of the share of a son. Subtract now one-third of what remains of the one-third, that is to say, one-ninth of the capital less one-seventh and one-third of the seventh of the share of a son; the remainder

---

* Since there are 3 sons, each son's share of the residue $=\frac{1}{3}$. Were there 3 sons and a daughter, the daughter's share would be $\frac{1}{7}$.

$$\tfrac{1}{3}-\tfrac{1}{7}=\tfrac{4}{7}$$

Let $x$ be the stranger's legacy, and $v$ a son's share

Then $1-x = 3v$

but $x = \tfrac{4}{7} v + \tfrac{1}{3}\left[\tfrac{1}{3}-\tfrac{4}{7} v\right]$

and $1-x = \tfrac{2}{3}+\tfrac{1}{3}-\tfrac{4}{7} v - \tfrac{1}{3}\left[\tfrac{1}{3}-\tfrac{4}{7} v\right] = 3v$

$\therefore \tfrac{2}{3}+\tfrac{2}{3}\left[\tfrac{1}{3}-\tfrac{4}{7} v\right] = 3v$

$\therefore \tfrac{2}{3}+\tfrac{2}{9} = 3\tfrac{8}{21} \times v$, or $\tfrac{8}{9} = \tfrac{71}{21} v$

$\therefore \tfrac{8}{3} = \tfrac{71}{7} v$  $\therefore v = \tfrac{56}{213} =$ a son's share

$x = \tfrac{45}{213} =$ the stranger's legacy.

( 99 )

is two-ninths of the capital less two sevenths and two-thirds of a seventh of the share of a son. Add this to the two-thirds of the capital; the sum is eight ninths of the capital less two-sevenths and two thirds of a seventh of the share of a son, or eight twenty-one parts of that share, and this is equal to three shares. Reduce this, you have then eight-ninths of the capital, equal to three shares and eight twenty-one parts of a share. Complete the capital by adding to eight-ninths as much as one-eighth of the same, and add in the same proportion to the shares. Then you find the capital equal to three shares and forty-five fifty-sixth parts of a share. Calculating now each share equal to fifty-six, the whole capital is two hundred and thirteen, the first legacy thirty-two, the second thirteen, and of the remaining one hundred and sixty-eight each son takes fifty-six.

*On another Species of Legacies.*

" A woman dies, leaving her daughter, her mother, and her husband, and bequeaths to some one as much as the share of her mother, and to another as much as one-ninth of her entire capital."\* Computation: You begin by dividing the heritage into thirteen parts, two

---

\* In the former examples (p. 90) when a husband and a mother were among the heirs, a husband was found to be entitled to $\frac{1}{4}=\frac{3}{12}$, and a mother to $\frac{1}{6}=\frac{2}{12}$ of the residue. Here a husband is stated to be entitled to $\frac{3}{13}$, and a mother to $\frac{2}{13}$ of the residue.

of which the mother receives. Now you perceive that the
(74) legacies amount to two parts plus one-ninth of the entire capital. Subtracting this, there remains eight-ninths of the capital less two parts, for distribution among the heirs. Complete the capital, by making the eight-ninths less two parts to be thirteen parts, and adding two parts to it, so that you have fifteen parts, equal to eight-ninths of capital; then add to this one-eighth of the same, and to the fifteen parts add likewise one-eighth of the same, namely, one part and seven-eighths; then you have sixteen parts and seven-eighths. The person to whom one-ninth is bequeathed, receives one-ninth of this, namely, one part and seven-eighths; the other, to whom as much as the share of the mother is bequeathed, receives two parts. The remaining thirteen parts are divided among the heirs, according to their legal shares. You best determine the respective shares by dividing the whole heritage into one hundred and thirty-five parts.

If she has bequeathed as much as the share of the husband and one-eighth and one-tenth of the capital,*

---

Let $\frac{1}{13}$ of the residue $= v$
$1 - \frac{1}{9} - 2v = 13v \quad \therefore \frac{8}{9} = 15v$
$\therefore v = \frac{8}{135}$ of the capital
A mother's share $= \frac{16}{135}$

* $\frac{1}{8} + \frac{1}{10} = \frac{9}{40}$

A husband's share of the residue is $\frac{3}{13}$
$\therefore 1 - \frac{9}{40} - 3v = 13v \quad \therefore \frac{31}{40} = 16v$
$\therefore v = \frac{31}{640}$; a husband's share $= \frac{93}{640}$
The stranger's legacy $= \frac{217}{640}$

( 101 )

then you begin by dividing the heritage into thirteen parts. Add to this as much as the share of the husband, namely, three; thus you have sixteen. This is what remains of the capital after the deduction of one-eighth and one-tenth, that is to say, of nine-fortieths. The remainder of the capital, after the deduction of one-eighth and one-tenth, is thirty-one fortieths of the same, which must be equal to sixteen parts. Complete your capital by adding to it nine thirty-one parts of the same, and multiply sixteen by thirty-one, which gives four hundred and ninety-six; add to this nine thirty-one parts of the same, which is one hundred and forty-four. The sum is six hundred and forty. Subtract one-eighth and one-tenth from it, which is one hundred and forty-four, and as much as the share of the husband, which is ninety-three. There remains four hundred and three, of which the husband receives ninety-three, the mother sixty-two, and every daughter one hundred and twenty-four.

(75)

If the heirs are the same,* but that she bequeaths to a person as much as the share of the husband, less one-ninth and one-tenth of what remains of the capital,

---

\* $\frac{1}{9}+\frac{1}{10}=\frac{19}{90}$

$1-3v+\frac{19}{90}[1-3v]=13v$

$\therefore \frac{109}{90}[1-3]=13v$

$\therefore \frac{109}{90}=[13+\frac{109}{30}]v$

$\therefore v=\frac{109}{1497}$

The husband's share $=\frac{327}{1497}$

The stranger's legacy $=\frac{80}{1497}$

after the subtraction of that share, the computation is this: Divide the heritage into thirteen parts. The legacy from the whole capital is three parts, after the subtraction of which there remains the capital less three parts. Now, one-ninth and one-tenth of the remaining capital must be added, namely, one-ninth and one-tenth of the whole capital less one-ninth and one-tenth of three parts, or less nineteen-thirtieths of a part; this yields the capital and one-ninth and one-tenth less three parts and nineteen-thirtieths of a part, equal to thirteen parts. Reduce this, by removing the three parts and nineteen-thirtieths from your capital, and adding them to the thirteen parts. Then you have the capital and one-ninth and one-tenth of the same, equal to sixteen parts and nineteen-thirtieths of a part. Reduce this to one capital, by subtracting from it nineteen one-hundred-and-ninths. There remains a

(76) capital, equal to thirteen parts and eighty one-hundred-and-ninths. Divide each part into one hundred and nine parts, by multiplying thirteen by one hundred and nine, and add eighty to it. This gives one thousand four hundred and ninety-seven parts. The share of the husband from it is three hundred and twenty-seven parts.

If some one leaves two sisters and a wife,* and bequeaths to another person as much as the share of a

---

* When the heirs are a wife, and 2 sisters, they each inherit $\frac{1}{3}$ of the residue.

Let

sister less one-eighth of what remains of the capital after the deduction of the legacy, the computation is this: You consider the heritage as consisting of twelve parts. Each sister receives one-third of what remains of the capital after the subtraction of the legacy; that is, of the capital less the legacy. You perceive that one-eighth of the remainder plus the legacy equals the share of a sister; and also, one-eighth of the remainder is as much as one-eighth of the whole capital less one-eighth of the legacy; and again, one-eighth of the capital less one-eighth of the legacy added to the legacy equals the share of a sister, namely, one-eighth of the capital and seven-eighths of the legacy. The whole capital is therefore equal to three-eighths of the capital plus three and five-eighth times the legacy. Subtract now from the capital three-eighths of the same. There remain five-eighths of the capital, equal to three and five-eighth times the legacy; and the entire capital is equal to five and four-fifth times the legacy. Consequently, if you assume the capital to be twenty-nine, the legacy is five, and each sister's share eight.

---

Let $x$ be the stranger's legacy.
$\frac{1}{3}[1-x] =$ a sister's share
$\frac{1}{3}[1-x] - \frac{1}{8}[1-x] = x$
$\therefore \frac{5}{24}[1-x] = x \quad \therefore \frac{5}{24} = \frac{29}{24}x$
$\therefore x = \frac{5}{29} \quad \therefore 1-x = \frac{24}{29}$
and a sister's share $= \frac{8}{29}$

( 104 )

### On another Species of Legacies.

"A man dies, and leaves four sons, and bequeaths to some person as much as the share of one of his sons; and to another, one-fourth of what remains after the deduction of the above share from one-third." You perceive that this legacy belongs to the class of those (77) which are taken from one-third of the capital.* Computation: Take one-third of the capital, and subtract from it the share of a son. The remainder is one-third of the capital less the share. Then subtract from it one-fourth of what remains of the one-third, namely, one-fourth of one-third less one-fourth of the share. The remainder is one-fourth of the capital less three-fourths of the share. Add hereto two-thirds of the capital: then you have eleven-twelfths of the capital less three-fourths of a share, equal to four shares. Reduce this by removing the three-fourths of the share from the capital, and adding them to the four shares. Then you have eleven-twelfths of the capital, equal to four shares and three-fourths. Complete your capital, by adding to the four shares and three-fourths one-fourth of the same. Then you have five shares and two-elevenths,

---

* Let the first bequest $= v$; and the second $= y$
Then $1 - v - y = 4v$
i.e. $\frac{2}{3} + \frac{1}{3} - v - \frac{1}{4}\left[\frac{1}{3} - v\right] = 4v$
$\therefore \frac{2}{3} + \frac{3}{4}\left[\frac{1}{3} - v\right] = 4v$
$\therefore \frac{2}{3} + \frac{3}{12} = [4 + \frac{3}{4}]v \quad \therefore \frac{11}{12} = \frac{19}{4}v$
$\therefore v = \frac{11}{57}$; the 2d bequest $= \frac{2}{57}$

equal to the capital. Suppose, now, every share to be eleven; then the whole square will be fifty-seven; one-third of this is nineteen; from this one share, namely, eleven, must be subtracted; there remain eight. The legatee, to whom one-fourth of this remainder was bequeathed, receives two. The remaining six are returned to the other two-thirds, which are thirty-eight. Their sum is forty-four, which is to be divided amongst the four sons; so that each son receives eleven.

If he leaves four sons, and bequeaths to a person as much as the share of a son, less one-fifth of what remains from one-third after the deduction of that share, then this is likewise a legacy, which is taken from one-third.* Take one-third, and subtract from it one share; there remains one-third less the share. Then return to it that which was excepted, namely, one-fifth of the one-third less one-fifth of the share. This gives one-third and one-fifth of one-third (or two-fifths) (78) less one share and one-fifth of a share. Add this to two-thirds of the capital. The sum is, the capital and one-third of one-fifth of the capital less one share and one-fifth of a share, equal to four shares. Reduce this by removing one share and one-fifth from the capital,

---

* $1 - v + \frac{1}{5} \left[ \frac{1}{3} - v \right] = 4v$
or $\frac{2}{3} \times \frac{1}{3} - v + \frac{1}{5} \left[ \frac{1}{3} - v \right] = 4v$
or $\frac{2}{3} + \frac{6}{5} \left[ \frac{1}{3} - v \right] = 4v$
$\therefore \frac{2}{3} + \frac{2}{5} = \left[ 4 + \frac{6}{5} \right] v \quad \therefore \frac{16}{15} = \frac{26}{5} v$
$\therefore v = \frac{8}{39}$, and the stranger's legacy $= \frac{7}{39}$

( 106 )

and add to it the four shares. Then you have the capital and one-third of one-fifth of the capital, which are equal to five shares and one-fifth. Reduce this to one capital, by subtracting from what you have the moiety of one-eighth of it, that is to say, one-sixteenth. Then you find the capital equal to four shares and seven-eighths of a share. Assume now thirty-nine as capital; one-third of it will be thirteen, and one share eight; what remains of one-third, after the deduction of that share, is five, and one-fifth of this is one. Subtract now the one, which was excepted from the legacy; the remaining legacy then is seven; subtracting this from the one-third of the capital, there remain six. Add this to the two-thirds of the capital, namely, to the twenty-six parts, the sum is thirty-two; which, when distributed among the four sons, yields eight for each of them.

If he leaves three sons and a daughter,* and bequeaths to some person as much as the share of a

---

* Since there are three sons and one daughter, the daughter receives $\frac{1}{7}$, and each son $\frac{2}{7}$ths of the residue.

If the 1st legacy $= v$, the 2d $= y$, and therefore a daughter's share $= v$,

$$1 - v - y = 7v; \quad \tfrac{1}{5} + \tfrac{1}{6} = \tfrac{11}{30}$$
$$\therefore \tfrac{5}{7} + \tfrac{2}{7} - v - \tfrac{11}{30}\left[\tfrac{2}{7} - v\right] = 7v$$
$$\text{i.e. } \tfrac{5}{7} + \tfrac{19}{30}\left[\tfrac{2}{7} - v\right] = 7v$$
$$\therefore \tfrac{5}{7} + \tfrac{19}{15 \times 7} = \left[7 + \tfrac{19}{30}\right]v$$
$$\therefore \tfrac{94}{7} = \tfrac{229}{2}v \quad \therefore = \tfrac{188}{1603}$$

The 2d legacy $= \ldots y = \tfrac{99}{1603}$

daughter, and to another one-fifth and one-sixth of what remains of two-sevenths of the capital after the deduction of the first legacy; then this legacy is to be taken out of two-sevenths of the capital. Subtract from two-sevenths the share of the daughter: there remain two-sevenths of the capital less that share. Deduct from this the second legacy, which comprises (79) one-fifth and one-sixth of this remainder: there remain one-seventh and four-fifteenths of one-seventh of the capital less nineteen-thirtieths of the share. Add to this the other five-sevenths of the capital: then you have six-sevenths and four-fifteenths of one-seventh of the capital less nineteen thirtieths of the share, equal to seven shares. Reduce this, by removing the nineteen thirtieths, and adding them to the seven shares: then you have six-sevenths and four-fifteenths of one-seventh of capital, equal to seven shares and nineteen-thirtieths. Complete your capital by adding to every thing that you have eleven ninety-fourths of the same; thus the capital will be equal to eight shares and ninety-nine one hundred and eighty-eighths. Assume now the capital to be one thousand six hundred and three; then the share of the daughter is one hundred and eighty-eight. Take two-sevenths of the capital; that is, four hundred and fifty-eight. Subtract from this the share, which is one hundred and eighty-eight; there remain two hundred and seventy. Remove one-fifth and one-sixth of this, namely, ninety-nine; the remainder is one hundred and seventy-one. Add thereto five-sevenths of the capital, which is one thousand one hundred and forty-five. The sum is one thousand three hundred and sixteen parts. This may be divided into seven shares, each of one hundred and eighty-eight parts; then this is the share of the daughter, whilst every son receives twice as much.

If the heirs are the same, and he bequeaths to some person as much as the share of the daughter, and to another person one-fourth and one-fifth out of what remains from two-fifths of his capital after the deduction of the share; this is the computation:* You must observe that the legacy is determined by the two fifths. Take two-fifths of the capital and subtract the shares: the remainder is, two-fifths of the capital less the share. Subtract from this remainder one-fourth and one-fifth of the same, namely, nine-twentieths of two-fifths, less as much of the share. The remainder is one-fifth and one-tenth of one-fifth of the capital less eleven-twentieths of the share. Add thereto three-fifths of the

---

\* $\frac{1}{4} + \frac{1}{5} = \frac{9}{20}$

Let the 1st legacy $= v =$ a daughter's share

Let the 2d legacy $= y$

$$1 - v - y = 7v$$

$$\therefore \tfrac{3}{5} + \tfrac{2}{5} - v - \tfrac{9}{20}\left[\tfrac{2}{5} - v\right] = 7v$$

$$\therefore \tfrac{3}{5} + \tfrac{11}{20}\left[\tfrac{2}{5} - v\right] = 7v$$

$$\therefore \tfrac{3}{5} + \tfrac{11}{10 \times 5} = \left[7 + \tfrac{11}{20}\right]v$$

$$\therefore \tfrac{41}{5} = \tfrac{151}{2}v \quad \therefore v = \tfrac{82}{755}$$

and the 2d legacy, $y, = \tfrac{99}{155}$

capital: the sum is four-fifths and one-tenth of one-fifth of the capital, less eleven-twentieths of the share, equal to seven shares. Reduce this by removing the eleven-twentieths of a share, and adding them to the seven shares. Then you have the same four-fifths and one-tenth of one-fifth of capital, equal to seven shares and eleven-twentieths. Complete the capital by adding to any thing that you have nine forty-one parts. Then you have capital equal to nine shares and seventeen eighty-seconds. Now assume each portion to consist of eighty-two parts; then you have seven hundred and fifty-five parts. Two-fifths of these are three hundred (81) and two. Subtract from this the share of the daughter, which is eighty-two; there remain two hundred and twenty. Subtract from this one-fourth and one-fifth, namely, ninety-nine parts. There remain one hundred and twenty-one. Add to this three-fifths of the capital, namely, four hundred and fifty-three. Then you have five hundred and seventy-four, to be divided into seven shares, each of eighty-two parts. This is the share of the daughter; each son receives twice as much.

If the heirs are the same, and he bequeaths to a person as much as the share of a son, less one-fourth and one-fifth of what remains of two-fifths (of the capital) after the deduction of the share; then you see that this legacy is likewise determined by two-fifths. Subtract two shares (of a daughter) from them, since every son receives two (such) shares; there remain

( 110 )

two-fifths of the capital less two (such) shares. Add thereto what was excepted from the legacy, namely, one-fourth and one-fifth of the two-fifths less nine-tenths of (a daughter's) share.* Then you have two-fifths and nine-tenths of one-fifth of the capital less two (daughter's) shares and nine-tenths. Add to this three-fifths of the capital. Then you have one capital and nine-tenths of one-fifth of the capital less two (daughter's) shares and nine-tenths, equal to seven (such) shares. Reduce this by removing the two shares and nine-tenths and adding them to the seven shares. Then you have one capital and nine-tenths of one-fifth of the capital, equal to nine shares of a daughter and nine-tenths. Reduce this to one entire capital, by deducting nine fifty-ninths from what you have. There remains the capital equal to eight such shares and twenty-three fifty-ninths. Assume now each share (of a daughter) to contain fifty-nine parts. Then the whole heritage comprizes four hundred and ninety-five parts. Two-fifths of this are one hundred and ninety-eight

---

\* $v = \frac{1}{7}$ of the residue = a daughter's share.

$2v$ = a son's share

$$1 - 2v + \tfrac{9}{20}\left[\tfrac{2}{5} - 2v\right] = 7v$$

i.e. $\tfrac{3}{5} + \tfrac{2}{5} - 2v + \tfrac{9}{20}\left[\tfrac{2}{5} - 2v\right] = 7v$

$\therefore \tfrac{3}{5} + \tfrac{29}{20}\left[\tfrac{2}{5} - 2v\right] = 7v$

$\therefore \tfrac{3}{5} + \tfrac{29}{10 \times} = \left[7 + \tfrac{29}{10}\right]v \quad \therefore \tfrac{59}{5} = 99v$

$\therefore v = \tfrac{59}{495}$; a son's share $= \tfrac{118}{495}$

and the legacy to the stranger $= \tfrac{82}{495}$

( 111 )

parts. Subtract therefrom the two shares (of a daughter) or one hundred and eighteen parts; there remain eighty parts. Subtract now that which was excepted, namely, one-fourth and one fifth of these eighty, or thirty-six parts; there remain for the legatee eighty-two parts. Deduct this from the parts in the total number of parts in the heritage, namely, four hundred and ninety-five. There remain four hundred and thirteen parts to be distributed into seven shares; the daughter receiving (one share or) fifty-nine (parts), and each son twice as much.

If he leaves two sons and two daughters, and bequeaths to some person as much as the share* of a

---

* Since there are two sons and two daughters, each son receives $\frac{1}{3}$, and each daughter $\frac{1}{6}$ of the residue. Let $v = $ a daughter's share.

$$\text{Let the 1st legacy } = x = v - \tfrac{1}{5}\left[\tfrac{1}{3} - v\right]$$
$$\ldots\ldots \text{ 2d } \ldots\ldots = y = v - \tfrac{1}{3}\left[\tfrac{1}{3} - x - v\right]$$
$$\text{and 3d } \ldots\ldots = \tfrac{1}{12}$$
$$1 - \tfrac{1}{12} - x - y = 6v$$
$$\text{i.e. } \tfrac{2}{3} - \tfrac{1}{12} + \tfrac{1}{3} - x - v + \tfrac{1}{3}\left[\tfrac{1}{3} - x - v\right] = 6v$$
$$\text{or } \tfrac{2}{3} - \tfrac{1}{12} + \tfrac{4}{3}\left[\tfrac{1}{3} - x - v\right] = 6v$$
$$\text{i.e. } \tfrac{7}{12} + \tfrac{4}{3}\left[\tfrac{1}{3} - v + \tfrac{1}{5}\left[\tfrac{1}{3} - v\right] - v\right] = 6v$$
$$\text{or } \tfrac{7}{12} + \tfrac{4}{3}\left[\tfrac{6}{5}\left[\tfrac{1}{3} - v\right] - v\right] 6v$$
$$\text{or } \tfrac{7}{12} + \tfrac{8}{15} = \left[6 + \tfrac{4 \times 11}{5 \times 5}\right]v = \tfrac{134}{15}v$$
$$\text{or } \tfrac{7}{4} + \tfrac{8}{5} = \tfrac{134}{5}a \quad \therefore v = \tfrac{67}{536} = \tfrac{1}{8}$$
$$\text{The 1st Legacy } = x = \tfrac{1}{12}$$
$$\text{The 2d } \ldots\ldots = y = \tfrac{1}{12}$$
$$\text{A son's share } = \tfrac{1}{4}$$

daughter less one-fifth of what remains from one-third after the deduction of that share; and to another person as much as the share of the other daughter less one-third of what remains from one-third after the deduction of all this; and to another person half one-sixth of his entire capital; then you observe that all these legacies are determined by the one-third. Take one-third of the capital, and subtract from it the share of a daughter; there remains one-third of the capital less one share. Add to this that which was excepted, namely, one-fifth of the one-third less one-fifth of the share: this gives one-third and one-fifth of one-third of (83) the capital less one and one-fifth portion. Subtract herefrom the portion of the second daughter; there remain one-third and one-fifth of one-third of the capital less two portions and one-fifth. Add to this that which was excepted; then you have one-third and three-fifths of one-third, less two portions and fourteen-fifteenths of a portion. Subtract herefrom half one-sixth of the entire capital: there remain twenty-seven sixtieths of the capital less the two shares and fourteen-fifteenths, which are to be subtracted. Add thereto two-thirds of the capital, and reduce it, by removing the shares which are to be subtracted, and adding them to the other shares. You have then one and seven-sixtieths of capital, equal to eight shares and fourteen-fifteenths. Reduce this to one capital by subtracting from every thing that you have seven-sixtieths. Then let a share be two hundred and one;* the whole capital will be one thousand six hundred and eight.

If the heirs are the same, and he bequeaths to a person as much as the share of a daughter, and one-fifth of what remains from one-third after the deduction of that share; and to another as much as the share of the second daughter and one-third of what remains from one-fourth after the deduction of that share: then, in the computation,† you must consider that the two legacies are determined by one-fourth and one-third. Take one-third of the capital, and subtract from it one share; there remains one-third of the capital less one share. Then subtract one-fifth of the remainder, namely, one-fifth of one-third of the capital, less one-fifth of the share; there remain four-fifths of one-third, less four-fifths of the share. Then take also one-fourth of the capital, and subtract from it one (84) share; there remains one-fourth of the capital, less one share. Subtract one-third of this remainder: there

---

\* $\frac{201}{1608} = \frac{1}{8} = \frac{3}{24} = v$; and $\frac{1}{12} = \frac{2}{24} = y$
The common denominator 1608 is unnecessarily great.

† Let $x$ be the 1st legacy; $y$ the 2d; $v$ a daughter's share.
$$1-x-y=6v$$
$$x=v+\tfrac{1}{5}\left[\tfrac{1}{3}-v\right]$$
$$y=v+\tfrac{1}{3}\left[\tfrac{1}{4}-v\right]$$
Then $1-\tfrac{1}{3}-\tfrac{1}{4}+\tfrac{1}{3}-v-\tfrac{1}{5}\left[\tfrac{1}{3}-v\right]+\tfrac{1}{4}-v-\tfrac{1}{3}\left[\tfrac{1}{4}-v\right]=6v$
or $\tfrac{5}{12}+\tfrac{4}{5}\left[\tfrac{1}{3}-v\right]+\tfrac{2}{3}\left[\tfrac{1}{4}-v\right]=6v$
$\therefore \tfrac{5}{12}+\tfrac{4}{15}+\tfrac{2}{12}=\left[6+\tfrac{4}{5}+\tfrac{2}{3}\right]v$
$\therefore \tfrac{51}{60}=\tfrac{112}{15}v \quad \therefore \tfrac{51}{448}=\tfrac{153}{1344}$
$x=\tfrac{212}{1344}; \; y=\tfrac{214}{1344}$

remain two-thirds of one-fourth of the capital, less two thirds of one share. Add this to the remainder from the one-third of the capital; the sum will be twenty-six sixtieths of the capital, less one share and twenty-eight sixtieths. Add thereto as much as remains of the capital after the deduction of one-third and one-fourth from it; that is to say, one-fourth and one-sixth; the sum is seventeen-twentieths of the capital, equal to seven shares and seven-fifteenths. Complete the capital, by adding to the portions which you have three-seventeenths of the same. Then you have one capital, equal to eight shares and one-hundred-and-twenty hundred-and-fifty-thirds. Assume now one share to consist of one-hundred-and-fifty-three parts, then the capital consists of one thousand three hundred and forty-four. The legacy determined by one-third, after the deduction of one share, is fifty-nine; and the legacy determined by one-fourth, after the deduction of the share, is sixty-one.

If he leaves six sons, and bequeaths to a person as much as the share of a son and one-fifth of what remains of one-fourth; and to another person as much as the share of another son less one-fourth of what remains of one-third, after the deduction of the two first legacies and the second share; the computation is this :*
You subtract one share from one-fourth of the capital;

---

\* Let $x$ be the legacy to the 1st stranger
and $y$ ............... 2d .......; $v =$ a son's share

there remains one-fourth less the share. Remove then (85) one-fifth of what remains of the one-fourth, namely, half one-tenth of the capital less one-fifth of the share. Then return to the one-third, and deduct from it half one-tenth of the capital, and four-fifths of a share, and one other share besides. The remainder then is one-third, less half one-tenth of the capital, and less one share and four-fifths. Add hereto one-fourth of the remainder, which was excepted, and assume the one-third to be eighty; subtracting from it half one-tenth of the capital, there remain of it sixty-eight less one share and four-fifths. Add to this one-fourth of it, namely, seventeen parts, less one-fourth of the shares to be subtracted from the parts. Then you have eighty-five parts less two shares and one fourth. Add this to the other two-thirds of the capital, namely, one hundred and sixty parts. Then you have one and one-eighth of one-sixth of capital, less two shares and one-fourth, equal to six shares. Reduce this, by removing the shares which are to be subtracted, and adding

---

$$1 - x - y = 6v$$
$$x = v + \tfrac{1}{5}[\tfrac{1}{4} - v] \, ; \, y = v - \tfrac{1}{4}[\tfrac{1}{3} - x - v]$$
i.e. $\tfrac{2}{3} + \tfrac{1}{3} - x - v + \tfrac{1}{4}[\tfrac{1}{3} - x - v] = 6v$
or $\tfrac{2}{3} + \tfrac{5}{4}[\tfrac{1}{3} - x - v] = 6v$
or $\tfrac{2}{3} + \tfrac{5}{4}[\tfrac{1}{3} - \tfrac{1}{4} + \tfrac{1}{4} - v - \tfrac{1}{5}[\tfrac{1}{4} - v] - v] = 6v$
or $\tfrac{2}{3} + \tfrac{5}{4}[\tfrac{1}{12} + \tfrac{4}{5}[\tfrac{1}{4} - v] - v] = 6v$
$\therefore \tfrac{2}{3} + \tfrac{5}{4 \times 12} + \tfrac{1}{4} = [7 + \tfrac{5}{4}]v$
$\therefore \tfrac{8}{3} + \tfrac{5}{12} + 1 = 33v \quad \therefore \tfrac{49}{12 \times 3 \times 3} = \tfrac{49}{396} = v$
$\therefore x = v + \tfrac{10}{396},$ and $y = v - \tfrac{6}{396}$

( 116 )

them to the other shares. Then you have one and one-eighth of one-sixth of capital, equal to eight shares and one-fourth. Reduce this to one capital, by subtracting from the parts as much as one forty-ninth of them. Then you have a capital equal to eight shares and four forty-ninths. Assume now every share to be forty-nine; then the entire capital will be three hundred and ninety-six; the share forty-nine; the legacy (86) determined by one-fourth, ten; and the exception from the second share will be six.

---

### On the Legacy with a Dirhem.

"A man dies, and leaves four sons, and bequeaths to some one a dirhem, and as much as the share of a son, and one-fourth of what remains from one-third after the deduction of that share." Computation:* Take

---

* Let the capital $= 1$; a dirhem $= \delta$;
the legacy $= x$; and a son's share $= v$
$$1 - x = 4v$$
$$x = v + \tfrac{1}{4}\left[\tfrac{1}{3} - v\right] + \delta$$
$$\therefore \tfrac{2}{3} + \tfrac{1}{3} - v - \tfrac{1}{4}\left[\tfrac{1}{3} - v\right] - \delta = 4v$$
$$\therefore \tfrac{2}{3} + \tfrac{3}{4}\left[\tfrac{1}{3} - v\right] - \delta = 4v$$
$$\therefore \tfrac{2}{3} + \tfrac{1}{4} - \delta = \left[4 + \tfrac{3}{4}\right]v$$
$$\therefore \tfrac{11}{12} - \delta = \tfrac{19}{4}v$$
$\therefore \tfrac{11}{57}$ of the capital $- \tfrac{12}{57}$ of a dirhem $= v$
and $\tfrac{13}{57}$ of the capital $+ \tfrac{48}{57}$ of a dirhem $= x$, the legacy.

If we assume the capital to be so many dirhems, or a dirhem to be such a part of the capital, we shall obtain the

one third of the capital and subtract from it one share; there remains one-third, less one share. Then subtract one-fourth of the remainder, namely, one-fourth of one-third, less one-fourth of the share; then subtract also one dirhem; there remain three-fourths of one-third of the capital, that is, one-fourth of the capital, less three-fourths of the share, and less one dirhem. Add this to two-thirds of the capital. The sum is eleven-twelfths of the capital, less three-fourths of the share and less one dirhem, equal to four shares. Reduce this by removing three-fourths of the share and one dirhem; then you have eleven-twelfths of the capital, equal to four shares and three-fourths, plus one dirhem. Complete your capital, by adding to the shares and one dirhem one-eleventh of the same. Then you have the capital equal to five shares and two-elevenths and one dirhem and one-eleventh. If you (87) wish to exhibit the dirhem distinctly, do not complete your capital, but subtract one from the eleven on account of the dirhem, and divide the remaining ten by the portions, which are four and three-fourths. The quotient is two and two-nineteenths of a dirhem. Assuming, then, the capital to be twelve dirhems, each

---

value of the son's share in terms of a dirhem, or of the capital only.

Thus, if we assume the capital to be 12 dirhems,
$v = \frac{12}{57}[11-1]\delta = \frac{120}{57}\delta = 2\frac{2}{19}$ dirhems,
$x = \frac{12}{57}[13+4]\delta = \frac{204}{57}\delta = 3\frac{11}{19}$ dirhems.

share will be two dirhems and two-nineteenths. Or, if you wish to exhibit the share distinctly, complete your square, and reduce it, when the dirhem will be eleven of the capital.

If he leaves five sons, and bequeaths to some person a dirhem, and as much as the share of one of the sons, and one-third of what remains from one-third, and again, one-fourth of what remains from the one-third after the deduction of this, and one dirhem more; then the computation is this:* You take one-third, and subtract one share; there remains one-third less one share. Subtract herefrom that which is still in your hands, namely, one-third of one-third less one-third of the share. Then subtract also the dirhem; there remain two-thirds of one-third, less two-thirds of the share and less one dirhem. Then subtract one-fourth of what you have, that is, one-eighteenth, less one-sixth of a share and less one-fourth of a dirhem, and

---

* Let the legacy $=x$; and a son's share $=v$

$$1-x=5v$$

$$\tfrac{2}{3}+\tfrac{1}{3}-v-\tfrac{1}{3}[\tfrac{1}{3}-v]-\delta-\tfrac{1}{4}[\tfrac{2}{3}[\tfrac{1}{3}-v]-\delta]-\delta=5v$$

i.e. $\tfrac{2}{3}+\tfrac{2}{3}[\tfrac{1}{3}-v]-\delta-\tfrac{1}{4}[\tfrac{2}{3}[\tfrac{1}{3}-v]-\delta]-\delta=5v$

i.e. $\tfrac{2}{3}+\tfrac{3}{4}[\tfrac{2}{3}[\tfrac{1}{3}-v]-\delta]-\delta=5v$

$\therefore \tfrac{2}{3}+\tfrac{1}{6}-\tfrac{1}{2}v-\tfrac{7}{4}\delta=5v$

$\therefore \tfrac{5}{6}-\tfrac{7}{4}\delta=\tfrac{11}{2}v$

$\therefore \tfrac{10}{66}$ of the capital $-\tfrac{21}{66}$ of a dirhem $=v$

$\therefore \tfrac{16}{66}$ of the capital $+\tfrac{105}{66}$ of a dirhem $=x$, the legacy.

If the capital $=\tfrac{45}{2}$ dirhems, or $\tfrac{1}{3}$ of the capital $=7\tfrac{1}{2}$ dirhems,

$v=\tfrac{34}{11}$ dirhems $=3\tfrac{1}{11}$ dirhems.

( 119 )

subtract also the second dirhem; the remainder is half one-third of the capital, less half a share and less one dirhem and three-fourths; add thereto two-thirds of the capital, the sum is five-sixths of the capital, less one half of a share, and less one dirhem and three-fourths, equal to five shares. Reduce this, by removing the (88) half share and the one dirhem and three-fourths, and adding them to the (five) shares. Then you have five-sixths of capital, equal to five shares and a half plus one dirhem and three-fourths. Complete your capital, by adding to five shares and a half and to one dirhem and three-fourths, as much as one-fifth of the same. Then you have the capital equal to six shares and three-fifths plus two dirhems and one-tenth. Assume, now, each share to consist of ten parts, and one dirhem likewise of ten; then the capital is eighty-seven parts. Or, if you wish to exhibit the dirhem distinctly, take the one-third, and subtract from it the share; there remains one-third, less one share. Assume the one-third (of the capital) to be seven and a half (dirhems). Subtract one-third of what you have, namely, one-third of one-third;* there remain two-thirds of one-third, less two-thirds of the share: that is, five dirhems, less two-thirds of the share. Then subtract one, on account of the one dirhem, and you retain four dirhems, less two-thirds

---

\* There is an omission here of the words " less one third of a share."

of the share. Subtract now one-fourth of what you have, namely, one part less one-sixth of a share; and remove also one part on account of the one dirhem; the remainder, then, is two parts less half a share. Add this to the two-thirds of the capital, which is fifteen (dirhems). Then you have seventeen parts less half a share, equal to five shares. Reduce this, by removing half a share, and adding it to the five shares. Then it is seventeen parts, equal to (89) five shares and a half. Divide now seventeen by five and a half; the quotient is the value of one share, namely, three dirhems and one-eleventh; and one-third (of the capital) is seven and a half (dirhems).

If he leaves four sons, and bequeaths to some person as much as the share of one of his sons, less one-fourth of what remains from one-third after the deduction of the share, and one dirhem; and to another one-third of what remains from the one-third, and one dirhem; then this legacy is determined by one-third.*

---

\* Let the 1st legacy be $x$, the 2d $y$; and a son's share $= v$

$$1 - x - y = 4v$$

i.e. $\frac{2}{3} + \frac{1}{3} - v + \frac{1}{4}[\frac{1}{3} - v] - \delta - \frac{1}{3}[\frac{1}{3} - v + \frac{1}{4}(\frac{1}{3} - v) - \delta] - \delta = 4v$

i.e. $\frac{2}{3} + \frac{2}{3}[\frac{1}{3} - v + \frac{1}{4}(\frac{1}{3} - v) - \delta] - \delta = 4v$

i.e. $\frac{2}{3} + \frac{2}{3}[\frac{5}{4}(\frac{1}{3} - v) - \delta] - \delta = 4v$

$\therefore \frac{2}{3} + \frac{5}{18} - \frac{5}{6}v - \frac{5}{3}\delta = 4v$

$\therefore \frac{17}{18} - \frac{5}{3}\delta = \frac{29}{6}v$

$\therefore \frac{17}{87} - \frac{20}{58}\delta = v$

also $\frac{14}{87} + \frac{33}{58}\delta = x$

$\frac{5}{87} + \frac{47}{58}\delta = y$

Take one-third of the capital, and subtract from it one share; there remains one-third, less one share; add hereto one-fourth of what you have: then it is one-third and one-fourth of one-third, less one share and one-fourth. Subtract one dirhem; there remains one-third of one and one-fourth, less one dirhem, and less one share and one-fourth. There remains from the one-third as much as five-eighteenths of the capital, less two-thirds of a dirhem, and less five-sixths of a share. Now subtract the second dirhem, and you retain five-eighteenths of the capital, less one dirhem and two-thirds, and less five-sixths of a share. Add to this two-thirds of the capital, and you have seventeen-eighteenths of the capital, less one dirhem and two-thirds, and less five-sixths of a share, equal to four shares. Reduce this, by removing the quantities which are to be subtracted, and adding them to the shares; then you have seventeen-eighteenths of the capital, equal to four portions and five-sixths plus one dirhem and two-thirds. Complete your capital by (90) adding to the four shares and five-sixths, and one dirhem and two-thirds, as much as one-seventeenth of the same. Assume, then, each share to be seventeen, and also one dirhem to be seventeen.* The whole capital will then be one hundred and seventeen. If you wish to exhibit the dirhem distinctly, proceed with it as I have shown you.

---

* Capital $=\frac{87}{17}v+\frac{30}{17}\delta$ ∴ if $v=17$, and $\delta=17$, capital $=117$

If he leaves three sons and two daughters, and bequeaths to some person as much as the share of a daughter plus one dirhem; and to another one-fifth of what remains from one-fourth after the deduction of the first legacy, plus one dirhem; and to a third person one-fourth of what remains from one-third after the deduction of all this, plus one dirhem; and to a fourth person one-eighth of the whole capital, requiring all the legacies to be paid off by the heirs generally: then you calculate this by exhibiting the dirhems distinctly, which is better in such a case.* Take one-fourth of the capital, and assume it to be six dirhems; the entire capital will be twenty-four dirhems. Subtract one share from the one-fourth; there remain six dirhems less one share. Subtract also one dirhem; there remain five dirhems less one share. Subtract

---

* Let the legacies to the three first legatees be, severally, $x$, $y$, $z$; the fourth legacy $= \frac{1}{8}$; and let a daughters' share $= v$.

$$\therefore \tfrac{7}{8} - x - y - z = 8v$$
$$x = v + \delta; \quad y = \tfrac{1}{5}\left[\tfrac{1}{4} - x\right] + \delta; \quad z = \tfrac{1}{4}\left[\tfrac{1}{3} - x - y\right] + \delta$$
Then $\tfrac{7}{8} - \tfrac{1}{3} + \tfrac{1}{3} - x - y - \tfrac{1}{4}\left[\tfrac{1}{3} - x - y\right] - \delta = 8v$
$$\therefore \tfrac{13}{24} + \tfrac{3}{4}\left[\tfrac{1}{3} - x - y\right] - \delta = 8v$$
but $\tfrac{1}{3} - x - y = \tfrac{1}{3} - \tfrac{1}{4} + \tfrac{1}{4} - x - \tfrac{1}{5}\left[\tfrac{1}{4} - x\right] - \delta$
$$= \tfrac{1}{12} + \tfrac{4}{5}\left[\tfrac{1}{4} - x\right] - \delta$$
$$= \tfrac{1}{12} + \tfrac{1}{5} - \tfrac{4}{5}v - \tfrac{9}{5}\delta$$
$$= \tfrac{17}{60} - \tfrac{4}{5}v - \tfrac{9}{5}\delta$$
$$\therefore \tfrac{13}{24} + \tfrac{3}{4} \times \tfrac{17}{60} - \tfrac{3}{5}v - \left[\tfrac{3}{4} \times \tfrac{9}{5} + 1\right]\delta = 8v$$
$\therefore \tfrac{181}{240} - \tfrac{47}{20}\delta = \tfrac{43}{5}v \quad \therefore v = \tfrac{181}{2064} - \tfrac{564}{2064}\delta$, and $1 = \tfrac{2064}{181}v + \tfrac{564}{181}\delta$
$x = \tfrac{181}{2064} + \tfrac{1500}{2064}\delta; \quad y = \tfrac{67}{2064} + \tfrac{1764}{2064}\delta; \quad z = \tfrac{110}{2064} + \tfrac{1248}{2064}\delta$

one-fifth of this remainder; there remain four dirhems, less four-fifths of a share. Now deduct the second dirhem, and you retain three dirhems, less four-fifths of a share. You know, therefore, that the legacy which was determined by one-fourth, is three dirhems, less four-fifths of a share. Return now to the one-third, which is eight, and subtract from it three dirhems, less four-fifths of a share. There remain five dirhems, less four-fifths of a share. Subtract also one-fourth of this and one dirhem, for the legacy; you then retain two dirhems and three-fourths, less three-fifths of a share. Take now one-eighth of the capital, namely, three; after the deduction of one-third, you retain one-fourth of a dirhem, less three-fifths of a share. Return now to the two-thirds, namely, sixteen, and subtract from them one-fourth of a dirhem less three-fifths of a share; there remain of the capital fifteen dirhems and three-fourths, less three-fifths of a share, which are equal to eight shares. Reduce this, by removing three-fifths of a share, and adding them to the shares, which are eight. Then you have fifteen dirhems and three-fourths, equal to eight shares and three-fifths. Make the division: the quotient is one share of the whole capital, which is twenty-four (dirhems). Every daughter receives one dirhem and one-hundred-and-forty-three one-hundred-and-seventy-second parts of a dirhem.*

---

* $v = \frac{181}{2064}$ of the capital $- \frac{564}{2064}$ of a dirhem. If we assume

If you prefer to produce the shares distinctly, take one-fourth of the capital, and subtract from it one share; there remains one-fourth of the capital less one share. Then subtract from this one dirhem: then subtract one-fifth of the remainder of one-fourth, which is one-fifth of one-fourth of the capital, less one-fifth of the share and less one-fifth of a dirhem; and subtract also the second dirhem. There remain four-fifths of the one-fourth less four-fifths of a share, and less one dirhem and four-fifths. The legacies paid out of one fourth amount to twelve two-hundred-and-fortieths of the capital and four-fifths of a share, and one dirhem and four-fifths. Take one-third, which is eighty, and subtract from it twelve, and four-fifths of a share, and one dirhem and four-fifths, and remove one-fourth of what remains, and one dirhem. You retain, then, of the one-third, only fifty-one, less three-fifths of a share, less two dirhems and seven-twentieths. Subtract herefrom one-eighth of the capital, which is thirty, and you retain twenty-one, less three-fifths of a share, and less two dirhems and seven-twentieths, and two-thirds of the capital, being equal to eight shares. Reduce this, by removing that which is to be subtracted, and adding it to the eight shares. Then you have one hundred and eighty-one parts of the

(92)

---

the capital to be equal to 24 dirhems

$$v = \frac{181 \times 24 - 564}{2064} \text{ dirhems} = \frac{4344 - 564}{2064} \delta$$
$$= \frac{3780}{2064} \delta = 1\frac{145}{172} \text{ dirhems.}$$

capital, equal to eight shares and three-fifths, plus two dirhems and seven twentieths. Complete your capital, by adding to that which you have fifty-nine one-hundred-and-eighty-one parts. Let, then, a share be three hundred and sixty two, and a dirhem likewise three hundred and sixty-two.* The whole capital is then five thousand two hundred and fifty-six, and the legacy out of one-fourth† is one thousand two hundred and four, and that out of one-third is four hundred and ninety-nine, and the one-eighth is six hundred and fifty-seven.

---

## On Completement.

"A woman dies and leaves eight daughters, a mo- (93) ther, and her husband, and bequeaths to some person as much as must be added to the share of a daughter to make it equal to one-fifth of the capital; and to another person as much as must be added to the share of the mother to make it equal to one-fourth of

---

\* The capital $= \frac{2064}{181}v + \frac{564}{181}\delta$

If we assume $v = 362$, and $\delta = 362$, the capital $= 5256$

Then $x = 724$; $y = 480$; $z = 499$; $\frac{1}{8}$th of capital $= 657$.

† The text ought to stand " the two first legacies are " instead of " the legacy out of one-fourth is."

The first legacy is .............. 724
The second ..................... 480
∴ the first + second legacy  =   1204

( 126 )

the capital."* Computation: Determine the parts of the residue, which in the present instance are thirteen. Take the capital, and subtract from it one-fifth of the same, less one part, as the share of a daughter: this being the first legacy. Then subtract also one-fourth, less two parts, as the share of the mother: this being the second legacy. There remain eleven-twentieths of the capital, which, when increased by three parts, are equal to thirteen parts. Remove now from thirteen parts the three parts on account of the three parts (on the other side), and you retain eleven-twentieths of the capital, equal to ten parts. Complete the capital, by adding to the ten parts as much as nine-elevenths of the same; then you find the capital equal to eighteen parts and two-elevenths. Assume now each part to be eleven; then the whole capital is two hundred, each part is eleven; the first legacy will be twenty-nine, and the second twenty-eight.

(94) If the case is the same, and she bequeaths to some person as much as must be added to the share of the husband to make it equal to one-third, and to another person as much as must be added to the share of the mother to make it equal to one-fourth; and to a

---

* In this case, the mother has $\frac{2}{13}$; and each daughter has $\frac{1}{13}$ of the residue.

$$1-x-y=13v$$
$$\text{i.e. } 1-\tfrac{1}{5}+v-\tfrac{1}{4}+2v=13v$$
$$\therefore \tfrac{11}{20}=10v \quad \therefore v=\tfrac{11}{200}; \quad x=\tfrac{29}{200}; \quad y=\tfrac{28}{200}$$

third as much as must be added to the share of a daughter to make it equal to one-fifth; all these legacies being imposed on the heirs generally: then you divide the residue into thirteen parts.* Take the capital, and subtract from it one-third, less three parts, being the share of the husband; and one-fourth, less two parts, being the share of the mother; and lastly, one-fifth less one part, being the share of a daughter. The remainder is thirteen-sixtieths of the capital, which, when increased by six parts, is equal to thirteen parts. Subtract the six from the thirteen parts: there remain thirteen-sixtieths of the capital, equal to seven parts. Complete your capital by multiplying the seven parts by four and eight-thirteenths, and you have a capital equal to thirty-two parts and four-thirteenths. Assuming then each part to be thirteen, the whole capital is four hundred and twenty.

If the case is the same, and she bequeaths to some person as much as must be added to the share of the mother to make it one-fourth of the capital; and to another as much as must be added to the portion of a daughter, to make it one-fifth of what remains of the capital, after the deduction of the first legacy; then

---

* $1-[\frac{1}{3}-3v]-[\frac{1}{4}-2v]-[\frac{1}{5}-v]=13v$
i.e. $1-\frac{1}{3}-\frac{1}{4}-\frac{1}{5}=7v$
$\therefore \frac{13}{60}=7v$
$\therefore v=\frac{13}{420}$

( 128 )

you constitute the parts of the residue by taking them out of thirteen.* Take the capital, and subtract from it one-fourth less two parts; and again, subtract one-fifth of what you retain of the capital, less one part; then look how much remains of the capital after the deduction of the parts. This remainder, namely, three-fifths of the capital, when increased by two parts and three-fifths, will be equal to thirteen parts. Subtract two parts and three-fifths from thirteen parts, there remain ten parts and two-fifths, equal to three-fifths of capital. Complete the capital, by adding to the parts which you have, as much as two-thirds of the same. Then you have a capital equal to seventeen parts and one-third. Assume a part to be three, then the capital is fifty-two, each part three; the first legacy will be seven, and the second six.

If the case is the same, and she bequeaths to some person as much as must be added to the share of the mother to make it one-fifth of the capital, and to another one-sixth of the remainder of the capital; then

---

\* $1-x-y=13v$

$x = \frac{1}{4} - 2v; \quad y = \frac{1}{5}[1-x] - v$

$1 - x - \frac{1}{5}[1-x] + v = 13v$

$\frac{4}{5}[1-x] = 12v \quad \therefore \quad \frac{4}{5}[\frac{3}{4} + 2v] = 12v$

$\therefore \quad \frac{3}{5} = [12 - \frac{8}{5}]v = \frac{52}{5}v$

$\therefore \quad v = \frac{3}{52} \quad \therefore \quad x = \frac{7}{52}, \quad y = \frac{6}{52}$

the parts are thirteen.* Take the capital, and subtract from it one-fifth less two parts; and again, subtract one-sixth of the remainder. You retain two-thirds of the capital, which, when increased by one part and two-thirds, are equal to thirteen parts. Subtract the one part and two-thirds from the thirteen parts: there remain two thirds of the capital, equal to eleven parts and one-third. Complete your capital, by adding to the parts as much as their moiety; thus you find the capital equal to seventeen parts. Assume now the capital to be eighty-five, and each part five; then the first legacy is seven, and the second thirteen, and the remaining sixty-five are for the heirs.

If the case is the same, and she bequeaths to some person as much as must be added to the share of the mother, to make it one-third of the capital, less that sum which must be added to make the share of a daughter equal to one-fourth of what remains of the capital after the deduction of the above complement; then the parts are thirteen.† Take the capital, and (96)

---

\* $1 - x - y = 13v$
$x = \frac{1}{5} - 2v; \quad y = \frac{1}{6}[1-x]$
$1 - x - \frac{1}{6}[1-x] = 13v$
$\therefore \frac{5}{6}[1-x] = 13v$
$\therefore \frac{5}{6}[\frac{4}{5} + 2v] = 13v$
$\therefore \frac{2}{3} + \frac{5}{3}v = 13v$
$\therefore \frac{2}{3} = \frac{3 \cdot 4}{3}v \quad \therefore v = \frac{1}{17}; \quad x = \frac{7}{85}; \quad y = \frac{13}{85}$

† $1 - x + y = 13v;$ and $x = \frac{1}{3} - 2v; \quad y = \frac{1}{4}[1-x] - v$
$\therefore 1 - x + \frac{1}{4}[1-x] - v = 13v$
$\therefore \frac{5}{4}[1-x] = 14v \quad \therefore \frac{5}{4}[\frac{2}{3} + 2v] = 14v$
$\therefore \frac{5}{6} = \frac{2 \cdot 3}{2}v \quad \therefore v = \frac{5}{69}; \quad x - y = \frac{4}{69}$

s

( 130 )

subtract from it one-third less two parts, and add to the remainder one-fourth (of such remainder) less one part; then you have five-sixths of the capital and one part and a half, equal to thirteen parts. Subtract one part and a half from thirteen parts. There remain eleven parts and a half, equal to five-sixths of the capital. Complete the capital, by adding to the parts as much as one-fifth of them. Thus you find the capital equal to thirteen parts and four-fifths. Assume, now, a part to be five, then the capital is sixty-nine, and the legacy four.

"A man dies, and leaves a son and five daughters, and bequeaths to some person as much as must be added to the share of the son to complete one-fifth and one-sixth, less one-fourth of what remains of one-third after the subtraction of the complement."* Take one-third of the capital, and subtract from it one-fifth and one-sixth of the capital, less two (seventh) parts; so that you retain two parts less four one hundred and twentieths of the capital. Then add it to the exception, which is half a part less one one hundred and

---

* Since there are five daughters and one son, each daughter receives $\frac{1}{7}$, and the son $\frac{2}{7}$ of the residue.

$$1-x=7v; \quad \tfrac{1}{5}+\tfrac{1}{6}=\tfrac{11}{30}$$
$$\therefore \tfrac{2}{3}+\tfrac{1}{3}-\tfrac{11}{30}+2v+\tfrac{1}{4}[\tfrac{1}{3}-\tfrac{11}{30}+2v] = 7v$$
$$\therefore \tfrac{2}{3}-\tfrac{1}{30}+2v+\tfrac{1}{4}[\tfrac{-1}{30}+2v] = 7v$$
$$\therefore \tfrac{2}{3}+\tfrac{5}{4}[\tfrac{-1}{30}+2v] = 7v$$
$$\therefore \tfrac{4}{6}-\tfrac{1}{24} = \tfrac{9}{2}v$$
$$\therefore \tfrac{5}{8} = \tfrac{9}{2}v \quad \therefore v=\tfrac{5}{36}, \text{ and } x=\tfrac{1}{36}$$

twentieth, and you have two parts and a half less five one hundred and twentieths of capital. Add hereto two-thirds of the capital, and you have seventy-five one hundred and twentieths of the capital and two parts and a half, equal to seven parts. Subtract, now, two parts and a half from seven, and you retain seventy-five one hundred and twentieths, or five-eighths, equal to four parts and a half. Complete your capital, by (97) adding to the parts as much as three-fifths of the same, and you find the capital equal to seven parts and one-fifth part. Let each part be five; the capital is then thirty-six, each portion five, and the legacy one.

If he leaves his mother, his wife, and four sisters, and bequeaths to a person as much as must be added to the shares of the wife and a sister, in order to make them equal to the moiety of the capital, less two-sevenths of the sum which remains from one-third after the deduction of that complement; the Computation is this :* If

---

* From the context it appears, that when the heirs of the residue are a mother, a wife, and 4 sisters, the residue is to be divided into 13 parts, of which the wife and one sister, together, take 5 : therefore the mother and 3 sisters, together, take 8 parts. Each sister, therefore, must take not less than $\frac{1}{13}$, nor more than $\frac{2}{13}$. In the case stated at page 102, a sister was made to inherit as much as a wife; in the present case that is not possible; but the widow must take not less than $\frac{3}{13}$; and each sister not more than $\frac{2}{13}$. Probably, in this case, the mother is supposed to inherit $\frac{2}{13}$; the wife $\frac{3}{13}$; each sister $\frac{2}{13}$.

you take the moiety from one-third, there remains one-sixth. This is the sum excepted. It is the share of the wife and the sister. Let it be five (thirteenth) parts. What remains of the one-third is five parts less one-sixth of the capital. The two-sevenths which he has excepted are two-sevenths of five parts less two-sevenths of one-sixth of the capital. Then you have six parts and three-sevenths, less one-sixth and two-sevenths of one-sixth of the capital. Add hereto two-thirds of the capital; then you have nineteen forty-seconds of the capital and six parts and three-sevenths, equal to thirteen parts. Subtract herefrom the six parts and three-sevenths. There remain nineteen forty-seconds of the capital, equal to six parts and four-sevenths. Complete your capital by adding to it its double and four-nineteenths of it. Then you find the capital equal to fourteen parts, and seventy

(98) one hundred and thirty-thirds of a part. Assume one part to be one hundred and thirty-three; then the whole capital is one thousand nine hundred and thirty-

---

$$x + 5v = \tfrac{1}{2}; \quad 1 - x + \tfrac{2}{7}\left[\tfrac{1}{3} - x\right] = 13v$$
$$\therefore \tfrac{2}{3} + \tfrac{1}{3} - x + \tfrac{2}{7}\left[\tfrac{1}{3} - x\right] = 13v$$
$$\therefore \tfrac{2}{3} + \tfrac{9}{7}\left[\tfrac{1}{3} - x\right] = 13v$$
$$\therefore \tfrac{2}{3} + \tfrac{9}{7}\left[\tfrac{-1}{6} + 5v\right] = 13v$$
$$\therefore \tfrac{2}{3} - \tfrac{3}{14} = \left[13 - \tfrac{45}{7}\right]v$$
$$\therefore \tfrac{19}{42} = \tfrac{46}{7}v \quad \therefore \tfrac{19}{276} = v$$
$$\therefore x = \tfrac{29}{276}, \text{ and the residue } = \tfrac{247}{276}$$

The author unnecessarily takes $7 \times 276 = 1932$ for the common denominator.

two; each part is one hundred and thirty-three, the completion of it is three hundred and one, and the exception of one-third is ninety-eight, so that the remaining legacy is two hundred and three. For the heirs remain one thousand seven hundred and twenty-nine.

## COMPUTATION OF RETURNS.*

### *On Marriage in Illness.*

" A man, in his last illness, marries a wife, paying (a marriage settlement of) one hundred dirhems, besides which he has no property, her dowry being

---

\* The solutions which the author has given of the remaining problems of this treatise, are, mathematically considered, for the most part incorrect. It is not that the problems, when once reduced into equations, are incorrectly worked out; but that in reducing them to equations, arbitrary assumptions are made, which are foreign or contradictory to the data first enounced, for the purpose, it should seem, of forcing the solutions to accord with the established rules of inheritance, as expounded by Arabian lawyers.

The object of the lawyers in their interpretations, and of the author in his solutions, seems to have been, to favour heirs and next of kin; by limiting the power of a testator, during illness, to bequeath property, or to emancipate slaves; and by requiring payment of heavy ransom for slaves whom a testator might, during illness, have directed to be emancipated.

ten dirhems. Then the wife dies, bequeathing one-third of her property. After this the husband dies."[*] Computation: You take from the one hundred that which belongs entirely to her, on account of the dowry, namely, ten dirhems; there remain ninety dirhems, out of which she has bequeathed a legacy. Call the sum given to her (by her husband, exclusive of her dowry) thing; subtracting it, there remain ninety dirhems less thing. Ten dirhems and thing are already in her hands; she has disposed of one-third of her property, which is three dirhems and one-third, and one-third of thing; there remain six dirhems and

---

[*] Let $s$ be the sum, including the dowry, paid by the man, as a marriage settlement; $d$ the dowry; $x$ the gift to the wife, which she is empowered to bequeath if she pleases.

She may bequeath, if she pleases, $d+x$; she actually does bequath $\frac{1}{3}[d+x]$; the residue is $\frac{2}{3}[d+x]$, of which one half, viz. $\frac{1}{3}[d+x]$ goes to her heirs, and the other half reverts to the husband

∴ the husband's heirs have $s-[d+x]+\frac{1}{3}[d+x]$ or $s-\frac{2}{3}[d+x]$; and since what the wife has disposed of, exclusive of the dowry, is $x$, twice which sum the husband is to receive, $s-\frac{2}{3}[d+x]=2x$ ∴ $\frac{1}{8}[3s-2d]=x$. But $s=100$; $d=10$ ∴ $x=35$; $d+x=45$; $\frac{1}{3}[d+x]=15$. Therefore the legacy which she bequeaths is 15, her husband receives 15, and her other heirs, 15. The husband's heirs receive $2x=70$.

But had the husband also bequeathed a legacy, then, as we shall see presently, the law would have defeated, in part, the woman's intentions.

two-thirds plus two-thirds of thing, the moiety of which, namely, three dirhems and one-third plus one-third of thing, returns as his portion to the husband.* Thus the heirs of the husband obtain (as his share) ninety-three dirhems and one-third, less two thirds of thing; and this is twice as much as the sum given to (99) the woman, which was thing, since the woman had power to bequeath one-third of all which the husband left;† and twice as much as the gift to her is two things. Remove now the ninety-three and one-third, from two-thirds of thing, and add these to the two things. Then you have ninety-three dirhems and one-third equal to two things and two-thirds. One thing is three-eighths of it, namely, as much as three-eighths of the ninety-three and one-third, that is, thirty-five dirhems.

If the question is the same, with this exception only, that the wife has ten dirhems of debts, and that she bequeaths one-third of her capital; then the Computa-

---

\* In other cases, as appears from pages 92 and 93, a husband inherits one-fourth of the residue of his wife's estate, after deducting the legacies which she may have bequeathed. But in this instance he inherits half the residue. If she die in debt, the debt is first to be deducted from her property, at least to the extent of her dowry (see the next problem.)

† When the husband makes a bequest to a stranger, the third is reduced to one-sixth. Vide p. 137.

tion is as follows:* Give to the wife the ten dirhems of her dowry, so that there remain ninety dirhems, out of which she bequeaths a legacy. Call the gift to her thing; there remain ninety less thing. At the disposal of the woman is therefore ten plus thing. From this her debts must be subtracted, which are ten dirhems. She retains then only thing. Of this she bequeaths one-third, namely, one-third of thing: there remains two-thirds of thing. Of this the husband receives by inheritance the moiety, namely, one-third of thing. The heirs of the husband obtain, therefore, ninety dirhems, less two-thirds of thing; and this is twice as much as the gift to her, which was thing; that is, two things. Reduce this, by removing the two-thirds of thing from ninety, and adding them to two things. Then you have ninety dirhems, equal to two things and two-thirds. One thing is three-eighths of this; that is to say, thirty-three dirhems and three-fourths, which is the gift (to the wife).

If he has married her, paying (a marriage settle-

---

* The same things being assumed as in the last example, $s-[d+x]$ remains with the husband; $d$ goes to pay the debts of the wife; and $\frac{x}{3}$ reverts from the wife to the husband.

$$\therefore s-d-\tfrac{2}{3}x = 2x \quad \therefore \tfrac{3}{8}[s-d] = x$$

$\therefore$ if $s = 100$, and $d = 10$, $x = 33\tfrac{3}{4}$; she bequeaths $11\tfrac{1}{4}$; $11\tfrac{1}{4}$ reverts to her husband; and her other heirs receive $11\tfrac{1}{4}$. The husband's heirs receive $2x = 67\tfrac{1}{2}$.

ment of one hundred dirhems, her dowry being ten (100) dirhems, and he bequeaths to some person one-third of his property; then the computation is this:* Pay to the woman her dowry, that is, ten dirhems; there remain ninety dirhems. Herefrom pay the gift to her, thing; then pay likewise to the legatee who is to receive one-third, thing: for the one-third is divided

---

* This case is distinguished from that in page 133 by two circumstances; first, that the woman does not make any bequest; second, that the husband bequeaths one-third of his property.

Suppose the husband not to make any bequest. Then, since the woman had at her disposal $d+x$, but did not make any bequest, $\frac{1}{2}[d+x]$ reverts to her husband; and the like amount goes to her other heirs.

$\therefore s-[d+x]+\frac{1}{2}[d+x] = 2x \quad \therefore x = \frac{1}{5}[2s-d]$

and since $s = 100$, and $d = 10$; $x = 38$; $d+x = 48$; $\frac{1}{2}[d+x] = 24$ reverts to the husband, and the like sum goes to her other heirs; and $2x = 76$, belongs to the husband's heirs.

Now suppose the husband to bequeath one-third of his property. The law here interferes with the testator's right of bequeathing; and provides that whatever sum is at the disposal of the wife, the same sum shall be at the disposal of the husband; and that the sum to be retained by the husband's heirs shall be twice the sum which the husband and wife together may dispose of.

$\therefore s - \frac{1}{2}[d+x] - x = 4x$

$\therefore \frac{1}{11}[2s-d] = x$; if $s = 100$, and $d = 10$; $x = \frac{190}{11} = 17\frac{3}{11}$; $d+x = 27\frac{3}{11}$; $\frac{1}{2}[d+x] = 13\frac{7}{11}$ reverts to the husband, and the like sum goes to the other heirs of the woman; $17\frac{3}{11}$ is what the husband bequeaths; and $69\frac{1}{11} = 4x$ goes to the husband's heirs.

into two moieties between them, since the wife cannot take any thing, unless the husband takes the same. Therefore give, likewise, to the legatee who is to have one-third, thing. Then return to the heirs of the husband. His inheritance from the woman is five dirhems and half a thing. There remains for the heirs of the husband ninety-five less one thing and a half, which is equal to four things. Reduce this, by removing one thing and a half, and adding it to the four things. There remain ninety-five, equal to five things and a half. Make them all moieties; there will be eleven moieties; and one thing will be equal to seventeen dirhems and three-elevenths, and this will be the legacy.

" A man has married a wife paying (a marriage settlement of) one hundred dirhems, her dowry being ten dirhems; and she dies before him, leaving ten dirhems, and bequeathing one-third of her capital; afterwards the husband dies, leaving one hundred and twenty dirhems, and bequeathing to some person one-third of his capital." Computation :* Give to the wife her dowry,

---

\* Let $c$ be the property which the wife leaves, besides $d$ the dowry, and $x$ the gift from the husband. She bequeaths $\frac{1}{3}[c+d+x]$; $\frac{1}{3}[c+d+x]$ goes to her husband; and $\frac{1}{3}[c+d+x]$ to her other heirs. The husband leaves property $s$, out of which must be paid the dowry, $d$; the gift to the wife, $x$; and the bequest he makes to the stranger, $x$; and his heirs receive from the wife's heirs $\frac{1}{3}[c+d+x]$

namely, ten dirhems; then one hundred and ten dirhems remain for the heirs of the husband. From these the (101) gift to the wife is thing, so that there remain one hundred and ten dirhems less thing; and the heirs of the woman obtain twenty dirhems plus thing. She bequeaths one-third of this, namely, six dirhems and two-thirds, and one-third of thing. The moiety of the residue, namely, six dirhems and two-thirds plus one-third of thing, returns to the heirs of the husband: so that one hundred and sixteen and two-thirds, less two-thirds of thing, come into their hands. He has bequeathed one-third of this, which is thing. There remain, therefore, one hundred and sixteen dirhems and two-thirds less one thing and two-thirds, and this is twice as much as the husband's gift to the wife added to his legacy to the stranger, namely, four things. Reduce this, and you find one hundred and sixteen dirhems and two-thirds, equal to five things and two-thirds. Consequently one thing is equal to

---

$s - d - 2x + \frac{1}{3}[c+d+x] = 4x$, according to the law of inheritance.

$$\therefore 3s + c - 2d = 17x, \text{ and } x = \frac{3s+c-2d}{17}$$

If $s = 120$, $c = 10$, and $d = 10$, $x = \frac{350}{17} = 20\frac{10}{17}$

$c+d+x = 40\frac{10}{17}$; $\frac{1}{3}[c+d+x] = 13\frac{9}{17}$

The wife bequeaths $13\frac{9}{17}$; $13\frac{9}{17}$ go to her husband, and $13\frac{9}{17}$ to her other heirs.

The husband bequeaths to the stranger $20\frac{10}{17}$; he gives the same sum to the wife; and $4x = 82\frac{6}{17}$ go to his heirs.

twenty dirhems and ten-seventeenths; and this is the legacy.

### On Emancipation in Illness.

"Suppose that a man on his death-bed were to emancipate two slaves; the master himself leaving a son and a daughter. Then one of the two slaves dies, leaving a daughter and property to a greater amount than his price.*" You take two-thirds of his price, and what the other slave has to return (in order to complete his (102) ransom). If the slave die before the master, then the son and the daughter of the latter partake of the heritage, in such proportion, that the son receives as much as the two daughters together. But if the slave die after the master, then you take two-thirds of his value and what is returned by the other slave, and distribute

---

* From the property of the slave, who dies, is to be deducted and paid to the master's heirs, first, two-thirds of the original cost of that slave, and secondly what is wanting to complete the ransom of the other slave. Call the amount of these two sums $p$; and the property which the slave leaves $\alpha$.

Next, as to the residue of the slaves' property:

First. If the slave dies before the master, the master's son takes $\frac{1}{2}[\alpha-p]$; the master's daughter $\frac{1}{4}[\alpha-p]$, and the slave's daughter $\frac{1}{4}[\alpha-p]$.

Second. If the slave dies after the master; the master's son is to receive $\frac{2}{3}p$, and the master's daughter $\frac{1}{3}p$; and then the master's son takes $\frac{1}{2}[\alpha-p]$, and the slave's daughter $\frac{1}{2}[\alpha-p]$.

it between the son and the daughter (of the master), in such a manner, that the son receives twice as much as the daughter; and what then remains (from the heritage of the slave) is for the son alone, exclusive of the daughter; for the moiety of the heritage of the slave descends to the daughter of the slave, and the other moiety, according to the law of succession, to the son of the master, and there is nothing for the daughter (of the master).

It is the same, if a man on his death-bed emancipates a slave, besides whom he has no capital, and then the slave dies before his master.

If a man in his illness emancipates a slave, besides whom he possesses nothing, then that slave must ransom himself by two-thirds of his price. If the master has anticipated these two-thirds of his price and has spent them, then, upon the death of the master, the slave must pay two-thirds of what he retains.* But if the master has anticipated from him his whole price and spent it, then there is no claim against the slave, since he has already paid his entire price.

" Suppose that a man on his death-bed emancipates a slave, whose price is three hundred dirhems, not having any property besides; then the slave dies, leaving three hundred dirhems and a daughter." The

---

* The slave retains one-third of his price; and this he must redeem at two-thirds of its value; namely at $\frac{2}{3} \times \frac{1}{3} = \frac{2}{9}$ of his original price.

( 142 )

computation is this:* Call the legacy to the slave thing. He has to return the remainder of his price, after the deduction of the legacy, or three hundred less thing. This ransom, of three hundred less thing, belongs to the master. Now the slave dies, and leaves thing and a (103) daughter. She must receive the moiety of this, namely, one half of thing; and the master receives as much. Therefore the heirs of the master receive three hundred less half a thing, and this is twice as much as the legacy, which is thing, namely, two things. Reduce this by removing half a thing from the three hundred, and adding it to the two things. Then you have three hundred, equal to two things and a half. One thing is, therefore, as much as two-fifths of three hundred,

---

\* Let the slave's original cost be $a$; the property which he dies possessed of, $\alpha$; what the master bequeaths to the slave, in emancipating him, $x$. Then the net property which the slave dies possessed of is $\alpha+x-a$. $\frac{1}{2}[\alpha+x-a]$ belongs, by law, to the master; and $\frac{1}{2}[\alpha+x-a]$ to the slave's daughter. The master's heirs, therefore, receive the ransom, $a-x$, and the inheritance, $\frac{1}{2}[\alpha+x-a]$; that is, $\frac{1}{2}[\alpha+a-x]$; and on the same principle as the slave, when emancipated, is allowed to ransom himself at two-thirds of his cost, the law of the case is that 2 are to be taken, where 1 is given.

$\therefore \frac{1}{2}[\alpha+a-x]=2x \quad \therefore x=\frac{1}{5}[\alpha+a]$

The daughter's share of the inheritance $=\frac{1}{5}[3\alpha-2a]$
The master's heirs receive............. $\frac{2}{5}[\alpha+a]$

If, as in the example, $\alpha=a$, $x=\frac{2}{5}a$; the daughter's share $=\frac{1}{5}a$; the heirs of the master receive $\frac{4}{5}a$.

namely, one hundred and twenty. This is the legacy (to the slave,) and the ransom is one hundred and eighty.

"Some person on his sick-bed has emancipated a slave, whose price is three hundred dirhems; the slave then dies, leaving four hundred dirhems and ten dirhems of debt, and two daughters, and bequeathing to a person one-third of his capital; the master has twenty dirhems debts." The computation of this case is the following:* Call the legacy to the slave thing; his ransom is the remainder of his price, namely, three hundred less thing. But the slave, when dying, left four hundred dirhems; and out of this sum, his ransom, namely, three hundred less thing, is paid to the

---

* Let the slave's original cost $=a$; the property he dies possessed of $=\alpha$; the debt he owes $=\varepsilon$

He leaves two daughters, and bequeaths to a stranger one-third of his capital.

The master owes debts to the amount $\mu$; where $a=300$; $\alpha=400$; $\varepsilon=10$; $\mu=20$.

Let what the master gives to the slave, in emancipating him $=x$.

Slave's ransom $=a-x$; slave's property—slave's ransom $= \alpha+x-a$

Slave's property — ransom — debt $=\alpha+x-a-\varepsilon$
Legacy to stranger $=\frac{1}{3}[\alpha+x-a-\varepsilon]$
Residue.........$=\frac{2}{3}[\alpha+x-a-\varepsilon]$

The master, and each daughter, are, by law, severally entitled to $\frac{1}{3} \times \frac{2}{3}[\alpha+x-a-\varepsilon]$

The master's heirs receive altogether $a-x+\frac{2}{9}[\alpha+x-a-\varepsilon]$ or $\frac{7}{9}[a-x]+\frac{2}{9}[\alpha-\varepsilon]$, which, on the principle that 2

master, so that one hundred dirhems and thing remain in the hands of the slave's heirs. Herefrom are (first) subtracted the debts, namely, ten dirhems; there remain then ninety dirhems and thing. Of this he has bequeathed one-third, that is, thirty dirhems and one-third of thing; so that there remain for the heirs sixty dirhems and two-thirds of thing. Of this the two daughters receive two-thirds, namely, forty dirhems and four-ninths of thing, and the master (104) receives twenty dirhems and two-ninths of thing, so that the heirs of the master obtain three hundred and twenty dirhems less seven-ninths of thing. Of this the debts of the master must be deducted, namely, twenty dirhems; there remain then three hundred dirhems less

---

are to be taken for 1 given, ought to be made equal to $2x$.

But the author directs that the equation for determining $x$ be

$$\tfrac{7}{9}[a-x] + \tfrac{2}{3}[\alpha-\varepsilon] - \mu = 2x$$

$$\therefore x = \tfrac{1}{25}[7a + 2[\alpha-\varepsilon] - 9\mu] \qquad = 108$$

Hence the slave receives, the debts which he owes, $\varepsilon = 10$
+ the legacy to the stranger $= \tfrac{1}{25}[9[\alpha-\varepsilon] - 6a - 3\mu] = 66$
+ the inheritance of 1st daughter $= \tfrac{1}{25}[6[\alpha-\varepsilon] - 4a - 2\mu] = 44$
+ the inheritance of 2d daughter $= \tfrac{1}{25}[6[\alpha-\varepsilon] - 4a - 2\mu] = 44$

$$\text{Total} = \tfrac{1}{25}[21\alpha + 4\varepsilon - 14a - 7\mu] = 164$$

And the master takes $\mu + 2x = \tfrac{1}{25}[4\alpha - 4\varepsilon + 14a - 7\mu] = 236$

Had the slave died possessed of no property whatever, his ransom would have been 200.

His ransom, here stated, exclusive of the sum which the master inherits from him, or $a-x, = 192$.

seven-ninths of thing; and this sum is twice as much as the legacy of the slave, which was thing; or, it is equal to two things. Reduce this, by removing the seven-ninths of thing, and adding them to two things; there remain three hundred, equal to two things and seven-ninths. One thing is as much as nine twenty-fifths of eight hundred, which is one hundred and eight; and so much is the legacy to the slave.

If, on his sick-bed, he emancipates two slaves, besides whom he has no property, the price of each of them being three hundred dirhems; the master having anticipated and spent two-thirds of the price of one of them before he dies;* then only one-third of the price

---

\* Were there the first slave only, who has paid off two-thirds of his original cost, the master having spent the money, that slave would have to complete his ransom by paying two-ninths of his original cost, that is $66\frac{2}{3}$ (see page 141).

Were there the second slave only, who has paid off none of his original cost, he would have to ransom himself at two-thirds of his cost; that is by paying 200 (see also page 141).

The master's heirs, in the case described in the text, are entitled to receive the same amount from the two slaves jointly, *viz.* $266\frac{2}{3}$, as they would be entitled to receive, according to the rule of page 141, from the two slaves, separately; but the payment of the sum is differently distributed; the slave who has paid two-thirds of his ransom being required to pay one-ninth only of his original cost; and the slave who has paid no ransom, being required to pay two-thirds of his own cost, and one-ninth of the cost of the first slave.

of this slave, who has already paid off a part of his ransom, belongs to the master; and thus the master's capital is the entire price of the one who has paid off nothing of his ransom, and one-third of the price of the other who has paid part of it; the latter is one hundred dirhems; the other three hundred dirhems: one-third of the amount, namely, one hundred and thirty-three dirhems and one third, is divided into two moieties among them; so that each of them receives sixty-six dirhems and two-thirds. The first slave, who has already paid two-thirds of his ransom, pays thirty-three dirhems and one-third; for sixty-six dirhems and two-thirds out of the hundred belong to himself as a legacy, and what remains of the hundred he must return. The second slave has to return two hundred and thirty-three dirhems and one-third.

(105)

"Suppose that a man, in his illness, emancipates two slaves, the price of one of them being three hundred dirhems, and that of the other five hundred dirhems; the one for three hundred dirhems dies, leaving a daughter; then the master dies, leaving a daughter likewise; and the slave leaves property to the amount of four hundred dirhems. With how much must every one ransom himself?"* The computation is this: Call

---

* Let A. be the first slave; his original cost $a$; the property he dies possessed of $\alpha$; and let B. be the second slave; and his cost $b$.

the legacy to the first slave, whose price is three hundred dirhems, thing. His ransom is three hundred dirhems less thing. The legacy to the second slave of a price of five hundred dirhems is one thing and two-thirds, and his ransom five hundred dirhems less one thing and two-thirds (*viz.* his price being one and two-thirds times the price of the first slave, whose ransom was thing, he must pay one thing and two-thirds for

---

Let $x$ be that which the master gives to A. in emancipating him.

A.'s ransom is $a-x$; and his property, minus his ransom, is $a-a+x$.

A.'s daughter receives $\frac{1}{2}[\alpha-a+x]$, and the master's heirs receive $\frac{1}{2}[\alpha-a+x]$

Hence the master receives altogether from A.,
$$a-x+\tfrac{1}{2}[\alpha-a+x] = \tfrac{1}{2}[\alpha+a-x.]$$

B.'s ransom is $b - \dfrac{b}{a}x$

The master's heirs receive from A. and B. together $\frac{1}{2}[\alpha+a+2b] - \dfrac{1}{2a}[a+2b]x$; and this is to be made equal to twice the amount of the legacies to A. and B., that is,
$$\tfrac{1}{2}[\alpha+a+2b] - \dfrac{1}{2a}[a+2b]x = 2\dfrac{a+b}{a}x$$
$$\therefore x = a\,\dfrac{\alpha+a+2b}{5a+6b} = \dfrac{1700}{15} = 113\tfrac{1}{3}$$

The master's heirs receive from A., $\dfrac{2a[\alpha+a+b]+3\alpha b}{5a+6b} = 293\tfrac{1}{3}$

A.'s daughter receives $[a+b]\dfrac{3\alpha-2}{5a+6}a = 800 \times \dfrac{600}{4500} = 106\tfrac{2}{3}$

The legacy to B. is $b\,\dfrac{\alpha+a+2b}{5a+6b} = 188\tfrac{8}{9}$; his ransom is $b\,\dfrac{4a+4b-\alpha}{5a+6b} = 311\tfrac{1}{9}$

The master's heirs receive from A. and B. together $2[a+b]\dfrac{\alpha+a+2b}{5a+6b} = 604\tfrac{4}{9}$.

his ransom). Now the slave for three hundred dirhems dies, and leaves four hundred dirhems. Out of this his ransom is paid, namely, three hundred dirhems less thing; and in the hands of his heirs remain one hundred dirhems plus thing: his daughter receives the moiety of this, namely, fifty dirhems and half a thing; and what remains belongs to the heirs of the master, namely, fifty dirhems and half a thing. This is added to the three hundred less thing; the sum is three hundred and fifty less half a thing. Add thereto the ransom of the other, which is five hundred dirhems less one thing and two-thirds; thus, the heirs (106) of the master have obtained eight hundred and fifty dirhems less two things and one-sixth; and this is twice as much as the two legacies together, which were two things and two-thirds. Reduce this, and you have eight hundred and fifty dirhems, equal to seven things and a half. Make the equation; one thing will be equal to one hundred and thirteen dirhems and one-third. This is the legacy to the slave, whose price is three hundred dirhems. The legacy to the other slave is one and two-thirds times as much, namely, one hundred and eighty-eight dirhems and eight-ninths, and his ransom three hundred and eleven dirhems and one-ninth.

"Suppose that a man in his illness emancipates two slaves, the price of each of whom is three hundred dirhems; then one of them dies, leaving five hundred dirhems and a daughter; the master having left a son."

( 149 )

Computation :* Call the legacy to each of them thing; the ransom of each will be three hundred less thing; then take the inheritance of the deceased slave, which is five hundred dirhems, and subtract his ransom, which is three hundred less thing; the remainder of his inheritance will be two hundred plus thing. Of this, one hundred dirhems and half a thing return to the master by the law of succession, so that now altogether four hundred dirhems less a half thing are in the hands of the master's heirs. Take also the ransom of the other slave, namely, three hundred dirhems less thing; then the heirs of the master obtain seven hundred dir-

---

* The first slave is A.; his cost $a$; his property $\alpha$; he leaves a daughter.

The second slave is B.; his cost $b$.

Then (as in page 147) $\frac{1}{2}[\alpha-a+x]$ goes to the daughter; and $x = a \; \frac{\alpha+a+2b}{5a+6b}$

The daughter receives $[a+b] \; \frac{3\alpha-2a}{5a+6b}$

The master receives from A. $\frac{2a[a+\alpha+b]+3\alpha b}{5a+6b}$

and the master receives from A. and B. together $2[a+b] \; \frac{\alpha+a+2b}{5a+6b}$

But if $b=a$ .................... $x = \frac{1}{11}[\alpha+3a] = 127\frac{3}{11}$
The daughter receives ........... $\frac{2}{11}[3\alpha-2a] = 163\frac{7}{11}$
The master receives from A. ...... $\frac{1}{11}[5\alpha+4a] = 336\frac{4}{11}$
The master receives from B. ........ $\frac{1}{11}[8a-\alpha] = 172\frac{8}{11}$
The master receives from A and B. .. $\frac{4}{11}[\alpha+3a] = 509\frac{1}{11}$

If $b=0$,
The daughter receives $\frac{1}{5}[3\alpha-2a]$
The master ........ $\frac{2}{5}[\alpha+a]$, as in page 142.

hems less one thing and a half, and this is twice as much as the sum of the two legacies of both, namely (107) two things, consequently as much as four things. Remove from this the one thing and a half: you find seven hundred dirhems, equal to five things and a half. Make the equation. One thing will be one hundred and twenty-seven dirhems and three-elevenths.

"Suppose that a man in his illness emancipate a slave, whose price is three hundred dirhems, but who has already paid off to his master two hundred dirhems, which the latter has spent; then the slave dies before the death of the master, leaving a daughter and three hundred dirhems."* Computation: Take the property left by the slave, namely, the three hundred, and add thereto the two hundred, which the master has spent; this together makes five hundred dirhems. Subtract from this the ransom, which is three hundred less thing

---

* The slave A. dies before his master, and leaves a daughter. His cost is $a$, of which he has redeemed $\hat{a}$, which the master has spent; and he leaves property $\alpha$.

Then the daughter receives .. $\frac{1}{2} [\alpha + \hat{a} - a + x]$
The master receives altogether $\frac{1}{2} [\alpha + \hat{a} + a - x]$
The master's heirs receive.... $\frac{1}{2} [\alpha - \hat{a} + a - x]$
And $\frac{1}{2} [\alpha - \hat{a} + a - x] = 2x$ ∴ $x = \frac{1}{5} [\alpha - \hat{a} + a]$
Hence the daughter receives $\frac{1}{5} [3\alpha + 2\hat{a} - 2a] = 140$
The master's heirs ........ $\frac{1}{5} [2\alpha - 2\hat{a} + 2a] = 160$
The master receives, in toto, $\frac{1}{5} [2\alpha + 3\hat{a} + 2a] = 360$

If the slave had not advanced, or the master had not spent $\hat{a}$, the daughter would have received $\frac{1}{5} [3\alpha + 3\hat{a} - 2a] = 180$ and the master would have received $\frac{1}{5} [2\alpha + 2\hat{a} + 2a] = 320$.

(since his legacy is thing); there remain two hundred dirhems plus thing. The daughter receives the moiety of this, namely, one hundred dirhems plus half a thing; the other moiety, according to the laws of inheritance, returns to the heirs of the master, being likewise one hundred dirhems and half a thing. Of the three hundred dirhems less thing there remain only one hundred dirhems less thing for the heirs of the master, since two hundred are spent already. After the deduction of these two hundred which are spent, there remain with the heirs two hundred dirhems less half thing, and this is equal to the legacy of the slave taken twice; or the moiety of it, one hundred less one-fourth of thing, is equal to the legacy of the slave, which is thing. Remove from this the one-fourth of thing; then you have one hundred dirhems, equal to one thing and one-fourth. One thing is four-fifths of it, namely, eighty dirhems. This is the legacy; and the ransom is two hundred and twenty dirhems. Add the inheritance of the slave, which is three hundred, to two hundred, which are spent by the master. The sum is five hundred dirhems. The master has received the ransom of two hundred and twenty dirhems; and the moiety of the remaining two hundred and eighty, namely, one hundred and forty, is for the daughter. Take these from the inheritance of the slave, which is three hundred; there remain for the heirs one hundred and sixty dirhems, and this is twice as much as the legacy of the slave, which was thing.

" Suppose that a man in his illness emancipates a slave, whose price is three hundred dirhems, but who has already advanced to the master five hundred dirhems; then the slave dies before the death of his master, and leaves one thousand dirhems and a daughter. The master has two hundred dirhems debts."* Computation: Take the inheritance of the slave, which is one thousand dirhems, and the five hundred, which the master has spent. The ransom from this is three hundred less thing. There remain therefore twelve hundred plus thing. The moiety of this belongs to the daughter: it is six hundred dirhems plus half a thing. Subtract it from the property left by the slave, which was one

---

* A.'s price is $a$; he has advanced to his master $á$; he leaves property $α$. He dies before his master, and leaves a daughter.

The master's debts are $\mu$; $x$ is what A. receives, in being emancipated; $a-x$ is the ransom; $\frac{1}{2}[α+á-a+x]$ is what the daughter receives.

Then $α-\frac{1}{2}[α+á-a+x]$ is what remains to the master; and $α-\frac{1}{2}[α+á-a+x]-\mu$ is what remains to him, after paying his debts; and this is to be made equal to $2x$.

Whence $x=\frac{1}{5}[α+a-á-2\mu]$

Hence the daughter receives .... $\frac{1}{5}[3α-2a+2á-\mu]=640$
The mother receives,
  inclusive of the debt } ........ $\frac{1}{5}[2α+2a-2á+\mu]=360$
The master receives,
  exclusive of the debt } ...... $\frac{1}{5}[2α+[2a-2á-4\mu]=160$

If the mode given in page 142 had been followed, it would have given $x=\frac{1}{5}[α+a+á-2\mu]$
and the daughter's portion $=\frac{1}{5}[3α-2a+3á-\mu]=740$.

thousand dirhems: there remain four hundred dirhems less half thing. Subtract herefrom the debts of the master, namely, two hundred dirhems; there remain two hundred dirhems less half thing, which are equal to the legacy taken twice, which is thing; or equal to two things. Reduce this, by means of the half thing. Then you have two hundred dirhems, equal to two things and a half. Make the equation. You find one thing, equal to eighty dirhems; this is the legacy. Add now the property left by the slave to the sum which he has advanced to the master: this is fifteen hundred dirhems. Subtract the ransom, which is two hundred and twenty dirhems; there remain twelve hundred and eighty dirhems, of which the daughter receives the moiety, namely, six hundred and forty dirhems. Subtract this from the inheritance of the slave, which is one thousand dirhems: there remain three hundred and sixty dirhems. Subtract from this the debts of the master, namely, two hundred dirhems; there remain then one hundred and sixty dirhems for the heirs of the master, and this is twice as much as the legacy of the slave, which was thing.

"Suppose that a man on his sick-bed emancipates a slave, whose price is five hundred dirhems, but who has already paid off to him six hundred dirhems. The master has spent this sum, and has moreover three hundred dirhems of debts. Now the slave dies, leaving his mother and his master, and property to the amount of seventeen hundred and fifty dirhems, with two hundred

( 154 )

dirhems debts." Computation:* Take the property left by the slave, namely, seventeen hundred and fifty dirhems, and add to it what he has advanced to the master, namely, six hundred dirhems; the sum is two thousand three hundred and fifty dirhems. Subtract from this the debts, which are two hundred dirhems, and the ransom, which is five hundred dirhems less thing, since the legacy is thing; there remain then sixteen hundred and fifty dirhems plus thing. The mother receives herefrom one-third, namely, five hundred and fifty plus one-third of thing. Subtract now this and the debts, which are two hundred dirhems, from the actual inheritance of the slave, which is seventeen hundred and fifty; there remain one thousand dirhems less one-third of thing. Subtract from this the debts of the master, namely, three hundred

(110)

---

\* A. dies before his master, and leaves a mother. His price was $a$; he has redeemed $\acute{a}$, which the master has spent. The property he leaves is $\alpha$. He owes debts $\varepsilon$. The master owes debts $\mu$.

$\frac{1}{3}[\alpha + \acute{a} - a + x - \varepsilon]$ is the mother's.

$\alpha - \frac{1}{3}[\alpha + \acute{a} - a + x - \varepsilon] - \varepsilon$ is the master's.

$\alpha - \frac{1}{3}[\alpha + \acute{a} - a + x - \varepsilon] - \varepsilon - \mu = 2x =$ the master's, after paying his debts.

Hence .................$x = \frac{1}{7}[2\alpha + a - \acute{a} - 2\varepsilon - 3\mu] = 300$
Mother's...............$= \frac{1}{7}[3\alpha - 2a + 2\acute{a} - 3\varepsilon - \mu] = 650$
Master's, without $\mu$ ....$= \frac{1}{7}[4\alpha + 2a - 2\acute{a} - 4\varepsilon - 6\mu] = 600$
Mother's, with $\mu$ .......$= \frac{1}{7}[4\alpha + 2a - 2\acute{a} - 4\varepsilon + \mu] = 900$
A. receives, inclusive of $\varepsilon = \frac{1}{7}[3\alpha - 2a + 2\acute{a} + 4\varepsilon - \mu] = 850$.

dirhems; there remain seven hundred dirhems less one-third of thing. This is twice as much as the legacy of the slave, which is thing. Take the moiety: then three hundred and fifty less one-sixth of thing are equal to one thing. Reduce this, by means of the one-sixth of thing; then you have three hundred and fifty, equal to one thing and one-sixth. One thing will then be equal to six-sevenths of the three hundred and fifty, namely, three hundred dirhems; this is the legacy. Add now the property left by the slave to what the master has spent already; the sum is two thousand three hundred and fifty dirhems. Subtract herefrom the debts, namely, two hundred dirhems, and subtract also the ransom, which is as much as the price of the slave less the legacy, that is, two hundred dirhems; there remain nineteen hundred and fifty dirhems. The mother receives one-third of this, namely, six hundred and fifty dirhems. Subtract this and the debts, which are two hundred dirhems, from the property actually left by the slave, which was seventeen hundred and fifty dirhems; there remain nine hundred dirhems. Subtract from this the debts of the master, which are three hundred dirhems; there remain six hundred dirhems, which is twice as much as the legacy.

" Suppose that some one in his illness emancipates a slave, whose price is three hundred dirhems: then the slave dies, leaving a daughter and three hundred dirhems; then the daughter dies, leaving her husband and three hundred dirhems; then the master dies." Computation:* Take the property left by the slave, which is three hundred dirhems, and subtract the ransom, which (111) is three hundred less thing; there remains thing, one half of which belongs to the daughter, while the other half returns to the master. Add the portion of the daughter, which is half one thing, to her inheritance, which is three hundred; the sum is three hundred dirhems plus half a thing. The husband receives the moiety of this; the other moiety returns to the master, namely one hundred and fifty dirhems plus one-fourth of thing. All that the master has received is therefore four hundred and fifty less one-fourth of thing; and this is twice as much as the legacy; or the moiety of it is as much as

---

* A. is emancipated by his master, and then dies, leaving a daughter, who dies, leaving a husband. Then the master dies.

A.'s price $= a$; his property $\alpha$. What he receives from the master $= x$.

The daughter's property $= \delta$

A.'s ransom $= a - x$. The daughter inherits $\frac{1}{2}[\alpha - a + x]$, and $\frac{1}{2}[\alpha - a + x]$ goes to the master.

$\frac{1}{2}[\delta + \frac{1}{2}[\alpha - a + x]]$ goes to the daughter's husband
and $\frac{1}{2}[\delta + \frac{1}{2}[\alpha - a + x]]$ to the master.

Hence, according to the author, we are to make
$$a - x + \frac{1}{2}[\alpha - a + x] + \frac{1}{2}[\delta + \frac{1}{2}[\alpha - a + x]] = 2x$$
$$\therefore x = \frac{1}{9}[3\alpha + a + 2\delta] = 200$$
Daughter's share $= \frac{1}{9}[6\alpha - 4a + \delta] = 100$
Husband's .... $= \frac{1}{9}[3\alpha - 2a + 5\delta] = 200$
Master's ...... $= \frac{1}{9}[2\alpha + 6a + 4\delta] = 400$.

the legacy itself, namely, two hundred and twenty five dirhems less one eighth thing are equal to thing. Reduce this by means of one-eighth of thing, which you add to thing; then you have two hundred and twenty-five dirhems, equal to one thing and one-eighth. Make the equation: one thing is as much as eight-ninths of two hundred and twenty-five, namely, two hundred dirhems.

" Suppose that some one in his illness emancipates a slave, of the price of three hundred dirhems; the slave dies, leaving five hundred dirhems and a daughter, and bequeathing one-third of his property; then the daughter dies, leaving her mother, and bequeathing one-third of her property, and leaving three hundred dirhems." Computation:* Subtract from the property left

---

* A. is emancipated, and dies, leaving a daughter, and bequeathing one-third of his property to a stranger.

The daughter dies, leaving a mother, and bequeathing one-third of her property to a stranger.

A.'s price is $a$; his property is $\alpha$

The daughter's property is $\delta$.

A.'s ransom is $a-x$; $\alpha-a+x$ is his property, clear of ransom.

$\frac{1}{3}[\alpha-a+x]$ goes to the stranger; and the like amount to A.'s daughter, and to the master.

$\frac{1}{3}[3\delta+\alpha-a+x]$ is the property left by the daughter.

$\frac{1}{9}[3\delta+\alpha-a+x]$ is the bequest of the daughter to a stranger.

$\frac{2}{9}[3\delta+\alpha-a+x]$ is the residue, of which $\frac{1}{3}$d,

  viz. $\frac{2}{27}[3\delta+\alpha-a+x]$ is the mother's,

  and $\frac{4}{27}[3\delta+\alpha-a+x]$ is the master's;

by the slave his ransom, which is three hundred dirhems less thing; there remain two hundred dirhems plus thing. He has bequeathed one-third of his property, that is, sixty-six dirhems and two-thirds plus one-third of thing. According to the law of succession, (112) sixty-six dirhems and two-thirds and one-third of thing belong to the master, and as much to the daughter. Add this to the property left by her, which is three hundred dirhems: the sum is three hundred and sixty-six dirhems and two-thirds and one-third of thing. She has bequeathed one-third of her property, that is, one hundred and twenty-two dirhems and two-ninths and one-ninth of thing; and there remain two hundred and forty-four dirhems and four-ninths and two-ninths of thing. The mother receives one-third of this, namely, eighty-one dirhems and four-ninths and one-third of one-ninth of a dirhem plus two-thirds of one-ninth of thing. The remainder returns to the master; it is a hundred and sixty-two dirhems and eight-ninths and two-thirds of one-ninth of a dirhem plus one-ninth and one-third of one-ninth of thing, as his share of the heritage.

---

Hence, according to the author, we are to make
$$a - x + \tfrac{1}{3}[\alpha - a + x] + \tfrac{4}{27}[3\delta + \alpha - a + x] = 2x$$
Therefore .......... $x = \tfrac{1}{68}[13\alpha + 14a + 12\delta] = 210\tfrac{5}{17}$
The daughter's share.. $= \tfrac{1}{68}[27\alpha - 18a + 4\delta] = 136\tfrac{13}{17}$
The daughter's bequest $= \tfrac{1}{68}[9\alpha - 6a + 24\delta] = 145\tfrac{10}{17}$
The mother's share .... $= \tfrac{2}{68}[3\alpha - 2a + 8\delta] = 97\tfrac{1}{17}$
The master's........ $= \tfrac{2}{68}[13\alpha + 14a + 12\delta] = 420\tfrac{10}{17}$.

Thus the master's heirs have obtained five hundred and twenty-nine dirhems and seventeen twenty-sevenths of a dirhem less four-ninths and one-third of one-ninth of thing; and this is twice as much as the legacy, which is thing. Halve it: You have two hundred and sixty-four dirhems and twenty-two twenty-sevenths of a dirhem, less seven twenty-sevenths of thing. Reduce it by means of the seven twenty-sevenths which you add to the one thing. This gives one hundred and sixty-four dirhems and twenty-two twenty-sevenths, equal to one thing and seven twenty-sevenths of thing. Make the equation, and adjust it to one single thing, by subtracting from it as much as seven thirty-fourths of the same. Then one thing is equal to two hundred and ten dirhems and five-seventeenths; and this is the legacy.

"Suppose that a man in his illness emancipates a slave, whose price is one hundred dirhems, and makes to some one a present of a slave-girl, whose price is five hundred dirhems, her dowry being one hundred dirhems, and the receiver cohabits with her." Abu Hanifah says: The emancipation is the more important act, and must first be attended to.

Computation:* Take the price of the girl, which is

---

\* The price of the slave-girl being $a$; and what she receives on being emancipated $x$, her ransom is $a-x$.

If her dowry is $α$, he that receives her, takes $α+x$.

five hundred dirhems; and remember that the price of the slave is one hundred dirhems. Call the legacy of the donee thing. The emancipation of the slave, whose price is one hundred dirhems, has already taken place. He has bequeathed one thing to the donee. Add the dowry, which is one hundred dirhems less one-fifth thing. Then in the hands of the heirs are six hundred dirhems less one thing and one-fifth of thing. This is twice as much as one hundred dirhems and thing; the moiety of it is equal to the legacy of the two, namely, three hundred less three-fifths of thing. Reduce this by removing the three-fifths of thing from three hundred, and add the same to one thing. This gives three hundred dirhems, equal to one thing and three-fifths and one hundred dirhems. Subtract now from three hun-

---

Hence, according to the author, we are to make
$$a-x=2\,[\alpha+x]\ ;\ \ \text{whence}\ x=\frac{a-2\alpha}{3}$$
And her ransom is $\frac{2}{3}\,[a+\alpha]$

But if a male slave be at the same time emancipated by the master, the donee must pay the ransom of that slave. If his price was $b$, $b-\frac{b}{a}x$ is his ransom.

Hence, according to the author, we are to make the sum of the two ransoms, viz. $a-x+b-\frac{b}{a}x = 2[\alpha+x]$

$$\therefore\ a+b-2\alpha=[3+\frac{b}{a}]\,x\ \ \therefore\ x=a\,\frac{a+b-2\alpha}{3a+b}=125$$

The donee pays ransom, in respect of the slave-girl $(a-x)=375$ and he pays ransom for the male slave ........ $b-\frac{b}{a}x=75$.

dred the one hundred, on account of the other one hundred. There remain two hundred dirhems, equal to one thing and three-fifths. Make the equation with this. One thing will be five-eighths of what you have; (114) take therefore five-eighths of two hundred. It is one hundred and twenty-five. This is thing; it is the legacy to the person to whom he had presented the girl.

"Suppose that a man emancipates a slave of a price of one hundred dirhems, and makes to some person a present of a slave girl of the price of five hundred dirhems, her dowry being one hundred dirhems; the donee cohabits with her, and the donor bequeaths to some other person one-third of his property." According to the decision of Abu Hanifah, no more than one-third can be taken from the first owner of the slave-girl; and this one-third is to be divided into two equal parts between the legatee and the donee. Computation:* Take the price of the girl, which is five hundred dirhems. The legacy out of this is thing; so that the heirs obtain five hundred dirhems less thing; and the dowry is one hundred less one-fifth of thing; consequently they

---

* The same notation being used as in the last example, the equation for determining $x$, according to the author, is to be

$$a - x + b - \frac{b}{a} x - x = 2\left[\alpha + 2x\right]$$

$$\therefore x = \frac{a}{6a+b}\left[a + b - 2\alpha\right] = 64\tfrac{16}{31}.$$

obtain six hundred dirhems less one thing and one-fifth of thing. He bequeaths to some person one third of his capital, which is as much as the legacy of the person who has received the girl, namely, thing. Consequently there remain for the heirs six hundred less two things and one-fifth, and this is twice as much as both their legacies taken together, namely, the price of the slave plus the two things bequeathed as legacies. Halve it, and it will by itself be equal to these legacies: it is then three hundred less one and one-tenth of thing. Reduce this by means of the one and one-tenth of thing. Then you have three hundred, equal to three things and one-tenth, plus one hundred dirhems. Remove one hundred on occount of (the opposite) one hundred; there remain two hundred, equal to three things and one-tenth. Make now the reduction. One thing will be as much as thirty-one (115) parts of the sum of dirhems which you have; and just so much will be the legacy out of the two hundred; it is sixty-four dirhems and sixteen thirty-one parts.

" Suppose that some one emancipates a slave girl of the price of one hundred dirhems, and makes to some person a present of a slave girl, which is five hundred dirhems worth; the receiver cohabits with her, and her dowry is one hundred dirhems; the donor bequeaths to some other person as much as one-fourth of his capital." Abu Hanifah says: The master of the girl cannot be required to give up more than one-third, and the legatee, who is to receive one-fourth, must give up one-fourth. Computation :* The price of the girl is five hundred dirhems. The legacy out of this is thing; there remain five hundred dirhems less thing. The dowry is one hundred dirhems less one-fifth of thing; thus the heirs obtain six hundred dirhems less one and one-fifth of thing. Subtract now the legacy of the person to whom one-fourth has been bequeathed, namely, three-fourths of thing; for if one-third is thing then one-fourth is as much as three-fourths of the same.

There remain then six hundred dirhems less one thing and thirty-eight fortieths. This is equal to the legacy taken twice. The moiety of it is equal to the legacies by themselves, namely, three hundred dirhems less thirty-nine fortieths of thing. Reduce this by means of the latter fraction. Then you have three hun- (116) dred dirhems, equal to one hundred dirhems and two things and twenty-nine fortieths. Remove one hundred on account of the other one hundred. There remain two hundred dirhems, equal to two things and twenty-nine-fortieths. Make the equation. You will then find one thing to be equal to seventy-three dirhems and forty-three one-hundred-and-ninths dirhems.

---

* The same notation being used as in the two former examples, the equation for determining $x$, according to the author, is

$$a-x+b-\frac{b}{a}\ x-\tfrac{3}{4}x = 2\left[\alpha+1\tfrac{3}{4}x\right]$$

Whence $x = \dfrac{4a}{21a+4b}\left[a+b-2\alpha\right] = 73\tfrac{43}{109}$.

## On return of the Dowry.

"A MAN, in the illness before his death, makes to some one a present of a slave girl, besides whom he has no property. Then he dies. The slave girl is worth three hundred dirhems, and her dowry is one hundred dirhems. The man to whom she has been presented, cohabits with her." Computation:* Call the legacy of the person to whom the girl is presented, thing. Subtract this from the donation: there remain three hundred less thing. One-third of this difference returns to the donor on account of dowry (since the dowry is one-third of the price): this is one hundred dirhems less one-third of thing. The donor's heirs obtain, therefore, four hundred less one and one-third of thing, which is equal to twice the legacy, which is thing, or to two things. Transpose the one and one-third thing from the four hundred, and add it to the two things; then you have four hundred, equal to three things and one-third. One thing is, therefore, equal to three-tenths of it, or to one hundred and twenty dirhems, and this is the legacy.

---

* Let $a$ be the slave-girl's price — $\alpha$ her dowry.
Then, according to the author, we are to make
$$a - x + \alpha - \frac{\alpha}{a} x = 2x$$
Therefore $x = \frac{a}{3a + \alpha} [a + \alpha] = \frac{3}{10} \times 400 = 120$
The donee is to receive the girl's dowry, worth 400, for 280.

"Or, suppose that he, in his illness, has made a present of the slave girl, her price being three hundred, her dowry one hundred dirhems; and the donor dies, after having cohabited with her." Computation:* Call the legacy thing: the remainder is three hundred less thing. The donor having cohabited with her, the dowry remains with him, which is one-third of the legacy, since the dowry is one-third of the price, or one-third of thing. Thus the donor's heirs obtain three (117) hundred less one and one-third of thing, and this is twice as much as the legacy, which is thing, or equal to two things. Remove the one and one-third of thing, and add the same to the two things. Then you have three hundred, equal to three things and one-third. One thing is, therefore, three-tenths of it, namely ninety dirhems. This is the legacy.

If the case be the same, and both the donor and donee have cohabited with her; then the Computation

---

* If the donor has cohabited with the slave-girl, the donor's heirs are to retain the dowry, but must allow the donee, in addition to the legacy $x$, the further sum of $\frac{a}{a}x$;

The ransom is then $a-x-\frac{a}{a}x$, which according to the author is to be made equal to $2x$.

$$\text{Whence } x = \frac{a^2}{3x+a} = 90$$

The donee is to receive the girl, worth 300, for 210.

is this:* Call the legacy thing; the deduction is three hundred dirhems less thing. The donor has ceded the dowry to the donee by (the donee's) having cohabited with her: this amounts to one-third of thing: and the donee cedes one-third of the deduction, which is one hundred less one-third of thing. Thus, the donor's heirs obtain four hundred less one and two-thirds of thing, which is twice as much as the legacy. Reduce this, by separating the one and two-thirds of thing from four hundred, and add them to the two things. Then you have four hundred things, equal to three things and two-thirds. One thing of these is three-elevenths of four hundred; namely, one hundred and

---

* If the donor has previously cohabited with the slave-girl, it appears from the last example, that the donee is entitled to ransom her for $a - x - \dfrac{\alpha}{a} x$.

If the donee cohabits with the slave-girl, it appears from the last example but one, that he is entitled to redeem the dowry, $\alpha$, for $\alpha - \dfrac{\alpha}{a} x$

The redemption of the girl and dowry is
$$a - x - \dfrac{\alpha}{a} x + \alpha - \dfrac{\alpha}{a} x,$$
which, according to the author, is to be made equal to $2x$.

That is $a + \alpha - \dfrac{a + 2\alpha}{a} x = 2x$

Whence $x = \dfrac{a}{3a + 2\alpha} \times [a + \alpha] = 109\tfrac{1}{11}$

The donee is to receive the girl and dowry, worth 400, for $290\tfrac{10}{11}$.

nine dirhems and one-eleventh. This is the legacy. The deduction is one hundred and ninety dirhems and ten-elevenths. According to Abu Hanifah, you call the thing a legacy, and what is obtained on account of the dowry is likewise a legacy.

If the case be the same, but that the donor, having cohabited with her, has bequeathed one-third of his (118) capital, then Abu Hanifah says, that the one-third is halved between the donee and the legatee. Computation:* Call the legacy of the person to whom the slave-girl has been given, thing. After the deduction of it, there remain three hundred, less thing. Then take the dowry, which is one-third of thing; so that the donor retains three hundred less one and one-third of thing: the donee's legacy being, according to Abu Hanifah, one and one-third of thing; according to other lawyers, only thing. The legatee, to whom one-third is bequeathed, receives as much as the legacy of the donee, namely, one and one-third of thing. The donor thus retains three hundred, less two things and

---

* The second case is here solved in a different way.
$$a - x - \frac{\alpha}{a}x = 2\left[x + \frac{\alpha}{a}x\right]$$
$$\therefore x = \frac{a^2}{3[\alpha + a]}$$
This being halved between the legatee and donee becomes
$$\frac{a^2}{6[\alpha + a]} = 37\frac{1}{2}$$
The donee receives the girl, worth 300, for $262\frac{1}{2}$.

two-thirds—equal to twice the two legacies, which are two things and two-thirds. The moiety of this, namely, one hundred and fifty less one and one-third of thing, must, therefore, be equal to the two legacies. Reduce it, by removing one and one-third of thing, and adding the same to the two legacies (things). Then you find one hundred and fifty, equal to four things. One thing is one-fourth of this, namely, thirty-seven and a half.

If the case be, that both the receiver and the donor have cohabited with her, and the latter has disposed of one-third of his capital by way of legacy; then the computation,* according to Abu Hanifah, is, that you call the legacy thing. After the deduction of it, there remain three hundred less thing. Then the dowry is taken, which is one hundred less one-third of thing; so that there are four hundred dirhems less one and one-third of thing. The sum returned from the dowry is one-third of thing; and the legatee, who is to receive one-third, obtains as much as the legacy of the first, namely, thing and one-third of thing. Thus there

---

* According to the author's rule, which is purely arbitrary,

$$a - 2x + \alpha - \frac{3\alpha}{a}x = 4\left[1 + \frac{\alpha}{a}\right]x$$

Whence $x = a \frac{a+\alpha}{6a+7\alpha} = 48$

The donee will have to redeem the girl and dowry, worth 400, for 352.

remain four hundred dirhems less three things, equal to twice the legacy, namely, two things and two-thirds. (119) Reduce this, by means of the three things, and you find four hundred, equal to eight things and one-third. Make the equation with this: one thing will be forty-eight dirhems.

" Suppose that a man on his sick-bed makes to another a present of a slave-girl, worth three hundred dirhems, her dowry being one hundred dirhems; the donee cohabits with her, and afterwards, being also on his sick-bed, makes a present of her to the donor, and the latter cohabits with her. How much does he acquire by her, and how much is deducted?"* Com-

---

* We have here the only instance in the treatise of a simple equation, involving two unknown quantities. For what the donee receives is one unknown quantity; and what the donor receives back again from the donee, called by the author " part of thing," is the other unknown quantity.

Let what the donee receives $=x$, and what the donor receives $=y$.

Then, retaining the same notation as before, according to the author, the donee receives, on the whole

$$x - y - [\alpha - \frac{\alpha}{a}x] + \frac{\alpha}{a}[x-y] = 2y$$

and the donor receives, on the whole

$$a - x + y + [\alpha - \frac{\alpha}{a}x] - \frac{\alpha}{a}[x-y] = 2[x + \frac{\alpha}{a}[x-y]]$$

Whence $x = \frac{1}{2} \cdot \frac{a}{4a^2 + 5a\alpha - \alpha^2}[3a^2 + 3a\alpha - 2\alpha^2] = 102$

$y = \frac{1}{2} \cdot \frac{a}{4a^2 + 5a\alpha - \alpha^2}[a^2 - 2\alpha^2] = 21$

putation: Take the price, which is three hundred dirhems; the legacy from this is thing; there remain with the donor's heirs three hundred less thing; and the donee obtains thing. Now the donee gives to the donor part of thing: consequently, there remains only thing less part of thing for the donee. He returns to the donor one hundred less one-third of thing; but takes the dowry, which is one-third of thing, less one-third of part of thing. Thus he obtains one and two-thirds thing less one hundred dirhems and less one and one-third of part of thing. This is twice as much as part of thing; and the moiety of it is as much as part of thing, namely, five-sixths of thing less fifty dirhems and less two-thirds of part of thing. Reduce this by removing two-thirds of part of thing and fifty dirhems. Then you have five-sixths of thing, equal to one and two-thirds of part of thing plus fifty dirhems. Reduce this to one single part of thing, in order to know what the amount of it is. You effect this by taking three-fifths (120) of what you have. Then one part of thing plus thirty dirhems is equal to half a thing; and one-half thing less thirty dirhems is equal to part of thing, which is the legacy returning from the donee to the donor. Keep this in memory.

Then return to what has remained with the donor;

---

But the reasons for reducing the question to these two equations are not given by the author, and seem to depend on the dicta of the sages of the Arabian law.

this was three hundred less thing: hereto is now added the part of thing, or one-half thing less thirty dirhems. Thus he obtains two hundred and seventy less half one thing. He further takes the dowry, which is one hundred dirhems less one-third thing, but has to return a dowry, which is one-third of what remains of thing after the subtraction of part of thing, namely, one-sixth of thing and ten dirhems. Thus he retains three hundred and sixty less thing, which is twice as much as thing and the dowry, which he has returned. Halve it: then one hundred and eighty less one-half thing are equal to thing and that dowry. Reduce this, by removing one-half thing and adding it to the thing and the dowry: you find one hundred and eighty dirhems, equal to one thing and a half plus the dowry which he has returned, and which is one-sixth thing and ten dirhems. Remove these ten dirhems; there remain one hundred and seventy dirhems, equal to one and two-thirds things. Reduce this, in order to ascertain what the amount of one thing is, by taking three-fifths of what you have; you find that one hundred and two are equal to thing, which is the legacy from the donor to the donee: and the legacy from the donee to the donor is the moiety of this, less thirty dirhems, namely, twenty-one.

## On Surrender in Illness.

(121) " Suppose that a man, on his sick-bed, deliver to some one thirty dirhems in a measure of victuals, worth ten dirhems; he afterwards dies in his illness; then the receiver returns the measure and returns besides ten dirhems to the heirs of the deceased." Computation: He returns the measure, the value of which is ten dirhems, and places to the account of the deceased twenty dirhems; and the legacy out of the sum so placed is thing; thus the heirs obtain twenty less thing, and the measure. All this together is thirty dirhems less thing, equal to two things, or equal to twice the legacy. Reduce it by separating the thing from the thirty, and adding it to the two things. Then, thirty are equal to three things. Consequently, one thing must be one-third of it, namely, ten, and this is the sum which he obtains out of what he places to the account of the deceased.

" Suppose that some one on his sick-bed delivers to a person twenty dirhems in a measure worth fifty dirhems; he then repeals it while still on his sick bed, and dies after this. The receiver must, in this case, return four-ninths of the measure, and eleven dirhems and one-ninth."* Computation: You know that the

---

\* Let $a$ be the gift of money; and the value of the measure $m \times a$.

It appears from the context that the donee is to pay the heirs $\frac{2}{3}ma$.

price of the measure is two and a half times as much as the sum which the donor has given the donee in money; and whenever the donee returns anything from the money capital, he returns from the measure as much as two and a half times that amount. Take now from the measure as much as corresponds to one thing, that is, two things and a half, and add this to what remains from the twenty, namely, twenty less thing. Thus the heirs of the deceased obtain twenty dirhems and one (122) thing and a half. The moiety of this is the legacy, namely, ten dirhems and three-fourths of thing; and this is one-third of the capital, namely, sixteen dirhems and two-thirds. Remove now ten dirhems on account of the opposite ten; there remain six dirhems and two-thirds, equal to three-fourths of thing. Complete the thing, by adding to it as much as one-third of the same; and add to the six dirhems and two-thirds

---

It is arbitrary how he shall apportion this sum between the money capital and the measure.

If he pays on the money capital $p \cdot a$
and on the measure $\ldots\ldots\ldots\ldots q \cdot ma$
we have the equation $p \cdot a + q \cdot ma = \frac{2}{3} ma$
or $p\ \ + q\ m\ = \frac{2}{3} m$

The author assumes $p = \frac{m}{2} \cdot q$

Whence $q = \frac{4}{9}$, and $p = \frac{5}{9}$, and therefore the donee pays on the money capital.... $\frac{5}{9} a = 11\frac{1}{9}$
and on the measure .... $\frac{4}{9} ma = 22\frac{2}{9}$

Total .................... $33\frac{1}{3}$.

likewise one-third of the same, namely, two dirhems and two-ninths; this yields eight dirhems and eight-ninths, equal to thing. Observe now how much the eight dirhems and eight-ninths are of the money capital, which is twenty dirhems. You will find them to be four-ninths of the same. Take now four-ninths of the measure and also five-ninths of twenty. The value of four-ninths of the measure is twenty-two dirhems and two-ninths; and the five-ninths of the twenty are eleven dirhems and one-ninth. Thus the heirs obtain thirty-three dirhems and one-third, which is as much as two-thirds of the fifty dirhems.—God is the Most Wise!

# NOTES.

*Page 1, line 2-5.*

The neglected state of the manuscript, in which most diacritical points are wanting, makes me very doubtful whether I have correctly understood the author's meaning in several passages of his preface.

In the introductory lines, I have considered the words التي باداء ما افترض منها علي من يعبده من خلقه as an amplification of what might briefly have been expressed by التي باداءها "through the performance of which." I conceive the author to mean, that God has prescribed to man certain duties, (ان الله قد افترض علي الناس شيئا من باداء ما افترض) ,المحامد and that by performing these (&c. we express our thankfulness (نقع اسم الشكر) &c.

Since my translation was made, I have had the advantage of consulting Mr. SHAKESPEAR about this passage. He prefers to read تقع , تستوجب , and تومن instead of نقع , نستوجب , and نومن, and proposes to translate as follows: " Praise to God for his favours in that which is proper for him from among his laudable deeds, which in the performance of what he has rendered indispensible from (or by reason of) them on (the part of) whoever of his creatures worships him, gives the name of thanksgiving, and secures the increase, and preserves from deterioration."

The construction here assumed is evidently easier than that adopted by myself, in as far as the relative pronoun التي representing صاحبه, is made the subject of the three subsequent verbs تقع, &c., whilst my translation presumes a transition from the third person (as in ما هو اهله, and in من يعبده) to the first (as in تقع, &c.).

A marginal note in the manuscript explains the words " لعل تقديره ونومن صاحبه من الغير ونومن من الغير by The meaning may be: we preserve from change him who enjoys it," (viz. the divine bounty, taking صاحبه for صاحب نعم الله. The *change* here spoken of is the forfeiture of the divine mercy by bad actions; for " God does not *change* the mercy which he bestows on men, as long as they do not *change* that which is within themselves." بان الله لم يكث مغيرا نعمة انعمها علي قوم حتي يغيروا ما بانفسهم (*Coran*, *Sur.* VIII. v. 55, *ed. Hinck.*).

### Page 1, *line* 7.

[علي حين فترة من الرسل] See *Coran*, *Sur.* v. v. 22. *ed. Hinck.*

### Page 1, *line* 14, 15.

I am particularly doubtful whether I have correctly read and translated the words of the text from واحتسابا to وذكره. Instead of احتسابا للاجر I should have preferred احسانا

" للاخر "benefitting others," if the verb احسن could be construed with the preposition ل.

### Page 2, line 1.

To the words رجل سبق a marginal note is given in the manuscript, which is too much mutilated to be here transcribed, but which mentions the names of several authors who first wrote on certain branches of science, and concludes with asserting, that the author of the present treatise was the first that ever composed a book on Algebra.

### Page 2, line 4.

An interlinear note in the manuscript explains فلم سعته by جمع مفترقه.

### Page 2, line 10.

MOHAMMED gives no definition of the science which he intends to treat of, nor does he explain the words جبر *jebr*, and مقابلة *mokābalah*, by which he designates certain operations peculiar to the solution of equations, and which, combined, he repeatedly employs as an expression for this entire branch of mathematics. As the former of these words has, under various shapes, been introduced into the several languages of Europe, and is now universally used as the designation of an important division of mathematical science, I shall here subjoin a few remarks on its original sense, and on its use in Arabic mathematical works.

The verb جبر *jabar* of which the substantive جبر *jebr* is derived, properly signifies to restore something broken, especially to cure a fractured bone. It is thus used in the following passage from MOTANABBI (p. 143, 144, *ed. Calcutt.*)

يا من الوذ بــه فيمــا اوملـــه   ومن اعوذ به مـمـــا احــــــــادره
ومن توهمت ان البحر راحتــه   جودا وان عطاياه جواهــــــره
ارحم شباب فتى اودت بجدته   يد البلا وذوي في السجن ناضره
لا يجبر الناس عظما انت كاسره   ولا يهيضون عظما انت جابـــره

"O thou on whom I rely in whatever I hope, with whom I seek refuge from all that I dread; whose bounteous hand seems to me like the sea, as thy gifts are like its pearls: pity the youthfulness of one, whose prime has been wasted by the hand of adversity, and whose bloom has been stifled in the prison. Men will not heal a bone which thou hast broken, nor will they break one which thou hast healed."

Hence the Spanish and Portuguese expression *algebrista* for a person who heals fractures, or sets right a dislocated limb.

In mathematical language, the verb جبر means, to make perfect, or to complete any quantity that is incomplete or liable to a diminution; *i. e.* when applied to equations, to transpose negative quantities to the opposite side by changing their signs. The negative quantity thus removed is construed with the particle بـ: thus, if $x^2-6=23$ shall be changed into $x^2=29$, the direction is اجبر المال بالستة وزدها علي الثلثة والعشرين i. e. literally "Restore the square from (the deficiency occasioned to it by) the six, and add these to the twenty-three."

The verb جبر is not likewise used, when in an equation an integer is substituted for a fractional power of the unknown quantity: the proper expression for this is either the second or fourth conjugation of كمل, or the second of تمّ.

The word مقابلة *mokābalah* is a noun of action of the verb قبل to be in front of a thing, which in the third conjugation is used in a reciprocal sense of two objects being opposite one another or standing face to face; and in the transitive sense of putting two things face to face, of confronting or comparing two things with one another.

In mathematical language it is employed to express the comparison between positive and negative terms in a compound quantity, and the reduction subsequent to such comparison. Thus $100 + 10x - 10x + x^2$ is reduced to $100 + x^2$ بعد ان قابلنا به " after we have made a comparison."

When applied to equations, it signifies, to take away such quantities as are the same and equal on both sides. Thus the direction for reducing $x^2 + x = x^2 + 4$ to $x = 4$ will be expressed by قابل.

In either application the verb requires the preposition ب before a pronoun implying the entire equation or compound quantity, within which the comparison and subsequent reduction is to take place.

The verb قابل is not likewise used, when the reduction of an equation is to be performed by means of a division: the proper term for this operation being ردّ.

The mathematical application of the substantives جبر and مقابلة will appear from the following extracts.

1. A marginal note on one of the first leaves of the Oxford manuscript lays down the following distinction:

اما الجبر وهو اتمام كل شيء ناقص بما يتم من غير جنسه والمقابلة من المفاعلة وهو المواجهة ولهذا يقال للمصلي القبلة اذا واجهها فلما صار لهذا الحساب جزيل عمله جبر الناقص [بما] نقص منه وزيادة مثل ما جبر به الناقص علي الجنس المقابل لتقابل الزيادة مثلما جبر به الناقص وكثر الاستعمال في ذلك فسمي جبرًا ومقابلة لانه يجبر كل شيء بما نقص منه و تقابل الاجناس بعضها الي بعض . . . . . . وقد صارت المقابلة ايضا تعرف [عند] اهل الحساب حذف المقادير المتشابهة

"*Jebr* is the restoration of anything defective by means of what is complete of another kind. *Mokābalah*, a noun of action of the third conjugation, is the facing a thing: whence it is applied to one praying, who turns his face towards the *kiblah*. In this branch of calculation, the method commonly employed is the restoring of something defective in its deficiency, and the adding of an amount equal to this restoration to the other side, so as to make the completion (on the one side) and this addition (on the other side) to face (or to balance) one another. As this method is frequently resorted to, it has been named *jebr* and *mokābalah* (or Restoring and Balancing), since here every thing is made complete if it is deficient, and the opposite sides are made to balance one another. . . . . . . Mathematicians also take the word *mokābalah* in the sense of the removal of equal quantities (from both sides of an equation)."

According to the first part of this gloss, in reducing $x-5a=10a$ to $x=15a$, the substitution of $x$ in place of $x-5a$ would afford an instance of *jebr* or restoration, and the corresponding addition of $5a$ to $10a$, would be an example of *mokābalah* or balancing. From the following extracts it will be seen, that *mokābalah* is more generally taken in the sense stated last by the gloss.

2. HAJI KHALFA, in his bibliographical work (MS. of the British Museum, fol. 167, *recto\**.) gives the following explanation: ومعنى الجبر زيادة قدر ما نقص في الجملة المعادلة بالاستثناء في الجملة الاخرى ليتعادلا ومعنى المقابلة اسقاط الزايد من احدي الجملتين للتعادل " *Jebr* is the adding to one side what is negative on the other side of an equation owing to a subtraction, so as to equalize them. *Mokābalah* is the removal of what is positive from either sum, so as to make them equal."

A little farther on HAJI KHALFA gives further illustration of this by an example: كما في قولنا عشرة الا شيئا يعدل اربعة اشياء فالجبر رفع الاستثناء بان يزاد مثل المستثنى علي المستثنى منه فيجعل العشرة كاملة كانه يجبر نقصانه ويزاد مثل المستثنى علي عديله كزيادة الشيء في المثال بعد جبر العشرة علي اربعة اشياء حتى تصير خمسة فالمقابلة ان تنقص

---

\* This manuscript is apparently only an abridgement of HAJI KHALFA's work.

الاجناس من الطرفين بعدة واحدة قيل هي تقابل بعض
الاشياء ببعض علي المساوات كما في مثال المذكور اذا قوبلت
العشرة بالخمسة علي المساوات وسمي العلم بهذين العلمين
..... علم الجبر والمقابلة لكثرة وقوعها فيه "For instance if
we say : 'Ten less one thing equal to four things;' then
*jebr* is the removal of the subtraction, which is performed
by adding to the minuend an amount equal to the subtrahend: hereby the ten are made complete, that which
was defective in them being restored. An amount equal
to the subtrahend is then added to the other side of the
equation: as in the above instance, after the ten have been
made complete, one thing must be added to the four things,
which thus become five things. *Mokābalah* consists in
withdrawing the same amount from quantities of the same
kind on both sides of the equation; or as others say, it is
the balancing of certain things against others, so as to
equalize them. Thus, in the above example, the ten are
balanced against the five with a view to equalize them.
This science has therefore been called by the name of
these two rules, namely, the rule of *jebr* or restoration,
and of *mokābalah* or reduction, on account of the frequent use that is made of them."

3. The following is an extract from a treatise by ABU
ABDALLAH AL-HOSAIN BEN AHMED,* entitled, المقدمة

---

\* I have not been able to find any information about this writer. The copy of the work to which I refer is comprized in the same volume with MOHAMMED BEN MUSA's work in the Bodleian library. It bears no date.

الكافية في اصول الجبر و المقابلة or " A complete introduction to the elements of algebra."

باب تفسير الجبر والمقابلة ۞ اعلم ان الحساب انما سموا هذا النوع جبرا لانهم وضعوه علي معادلة ........ فلما كانوا وضعوه علي المعادلة اداهم العمل في اكثر مسائلة الي معادلة الناقص بغير الناقص فلم يكن بد من جبر ذلك الناقص بما ينقص وزيادة مثل ذلك علي ما عدله فلما كثر ذلك فيه سموه جبرا فهذا معني الجبر وعلة تسميتهم به هذا النوع ۞ فاما المقابلة فهو حذف المقادير المتشابهة من الجهتين ۞

"On the original meaning of the words *jebr* and *mokābalah*. This species of calculation is called *jebr* (or completion) because the question is first brought to an equation ...... And as, after the equation has been formed, the practice leads in most instances to equalize something defective with what is not defective, that defective quantity must be completed where it is defective; and an addition of the same amount must be made to what is equalized to it. As this operation is frequently employed (in this kind of calculation), it has been called *jebr*: such is the original meaning of this word, and such the reason why it has been applied to this kind of calculation. *Mokābalah* is the removal of equal magnitudes on both sides (of the equation)."

4. In the *Kholāset al Hisāb*, a compendium of arithmetic and geometry by BAHA-EDDIN MOHAMMED BEN AL HOSAIN (died A.H. 1031, i. e. 1575 A.D.) the Arabic

text of which, together with a Persian commentary by Roshan Ali, was printed at Calcutta* (1812. 8vo.) the following explanation is given (pp. 334. 335.) والطرف

وذ الاستثناء يكمل ويزاد مثل ذلك على الاخر وهو الجبر
والاجناس المتجانسة المستوية في الطرفين تسقط منهما وهو المقابلة

" The side (of the equation) on which something is to be subtracted, is made complete, and as much is added to the other side: this is *jebr*; again those cognate quantities which are equal on both sides are removed, and this is *mokābalah*." The examples which soon follow, and the solution of which Baha-eddin shows at full length, afford ample illustration of these definitions. In page 338, $1500 - \frac{1}{4}x = x$ is reduced to $1500 = 1\frac{1}{4}x$; this he says is effected by *jebr*. In page 341, $7x = \frac{1}{2}x^2 + \frac{1}{2}x$ is reduced to $13x = x^2$, and this he states to be the result of both *jebr* and *mokābalah*.

The Persians have borrowed the words *jebr* and *mokābalah*, together with the greater part of their mathematical terminology, from the Arabs. The following extract from a short treatise on Algebra in Persian verse, by Mohammed Nadjm-eddin Khan, appended to the Calcutta edition of the *Kholāset al Hisāb*, will serve as an illustration of this remark.

---

* A full account of this work by Mr. Strachey will be found in the twelfth volume of the Asiatic Researches, and in Hutton's Tracts on mathematical and philosophical subjects, vol. II. pp 179-193. See also Hutton's Mathematical Dictionary, art. *Algebra*.

طرفي كه دروست حرف الّا
تكميل كن و مثل آن را
بر طرف دگر فزون كن اي حبر
در مصطلح است نام اين جبر
هنگام معادله تو بشناس
افتد اگر اين كه بعض اجناس
با وصف تجانس از سويت
در هر طرف اند بي مزيت
بايد كه زهر دو سو براني
نامش تو مقابله بخواني

" Complete the side in which the expression *illā* (less, minus) occurs, and add as much to the other side, O learned man: this is in correct language called *jebr*. In making the equation mark this: it may happen that some terms are cognate and equal on each side, without distinction; these you must on both sides remove, and this you call *mokābalah*."

With the knowledge of Algebra, its Arabic name was introduced into Europe. LEONARDO BONACCI of Pisa, when beginning to treat of it in the third part of his treatise of arithmetic, says: *Incipit pars tertia de solutione quarundam quæstionum secundum modum Algebræ et Almucabalæ, scilicet oppositionis et restaurationis.* That the sense of the Arabic terms is here given in the inverted order, has been remarked by COSSALI. The definitions of *jebr* and *mokābalah* given by another early Italian writer, Lucas Paciolus, or Lucas de Burgo, are thus reported by Cossali: *Il commune oggetto dell' operar loro è recare la equazione alla sua maggior unità. Gli uffizj loro per questo commune intento sono contrarj: quello dell'* Algebra *è di restorare li extremi dei diminuti; e quello di* Almucabala *di levare da li extremi i superflui. Intende Fra Luca per* extremi *i membri dell' equazione.*

Since the commencement of the sixteenth century, the word *mokābalah* does no longer appear in the title of Algebraic works. Hieronymus Cardan's Latin treatise, first published in 1545, is inscribed: *Artis magnæ sive de regulis algebraicis liber unus.* A work by John Scheubelius, printed at Paris in 1552, is entitled: *Algebræ compendiosa facilisque descriptio, qua depromuntur magna Arithmetices miracula.* (See Hutton's Tracts, &c. II. pp. 241-243.) Pelletier's Algebra appeared at Paris in 1558, under the title: *De occulta parte numerorum quam Algebram vocant, libri duo.* (Hutton, l. c. p. 245. Montucla, *hist. des math.* I. p. 613.) A Portuguese treatise, by Pedro Nuñez or Nonius, printed at Amberez in 1567, is entitled: *Libro de Algebra y Arithmetica y Geometria.* (Montucla, l. c. p. 615.)

In Feizi's Persian translation of the *Lilavati* (written in 1587, printed for the first time at Calcutta in 1827, 8vo.) I do not recollect ever to have met with the word جبر; but مقابلة is several times used in the same sense as in the above Persian extract.

*Page* 3, *line* 3, *seqq.*

In the formation of the numerals, the thousand is not, like the ten and the hundred, multiplied by the units only, but likewise by any number of a higher order, such as tens and hundreds: there being no special words in Arabic (as is the case in Sanscrit) for ten-thousand, hundred-thousand, &c.

From this passage, and another on page 10, it would appear that our author uses the word عقد, *plur.* عقود, knot or tie, as a general expression for *all* numerals of a higher order than that of the units. Baron S. DE SACY, in his Arabic Grammar, (vol. I. § 741) when explaining the terms of Arabic grammar relative to numerals, translates عقود by *nœuds*, and remarks: *Ce sont les noms des dixaines, depuis* vingt *jusqu'*$^{à}$ quatre-vingt-dix.

*Page* 3, *line* 9-11.

The forms of algebraic expression employed by LEONARDO are thus reported by COSSALI (*Origine*, &c. *dell' Algebra*, I. p. 1.): *Tre considerazioni distingue Leonardo nel numero : una assoluta, o semplice, ed è quella del numero in se stesso ; le altre due relative, e sono quelle di radice e di quadrato. Nominando il quadrato soggiugne* QUI VIDELICET CENSUS DICITUR, *ed il nome di censo è quello di cui in seguito si serve.* That LEONARDO seems to have chosen the expression *census* on account of its acceptation, which is correspondent to that of the

Arabic مَال, has already been remarked by Mr. Cole-brooke (Algebra, &c., Dissertation, p. liv.)

Paciolo, who wrote in Italian, used the words *numero*, *cosa*, and *censo;* and this notation was retained by Tartaglia. From the term *cosa* for the unknown number, exactly corresponding in its acceptation to the Arabic شَيء thing, are derived the expressions *Ars cossica* and the German *die Coss*, both ancient names of the science of Algebra. Cardan's Latin terminology is *numerus, quadratum,* and *res,* for the latter also *positio* or *quantitas ignota.*

### Page 3, line 17.

I have added from conjecture the words عدد ادا وحذور تعدل which are not in the manuscript. There occur several instances of such omissions in the work.

The order in which our author treats of the simple equations is, 1st. $x^2 = px$; 2d. $x^2 = n$; 3d. $px = n$. Leonardo had them in the same order. (See Cossali, l. c. p. 2.) In the *Kholāset al Hisāb* the arrangement is, 1st. $n = px$; 2d. $px = x^2$; 3d. $n = x^2$.

### Page 5, line 9.

In the *Lilavati,* the rule for the solution of the case $cx^2 + bx = a$ is expressed in the following stanza.

गुणघ्नमूलोनयुतस्य राशे
दृष्टस्य युक्तस्य गुणार्धकृत्या १

मूलं गुणार्धेन युतं विहीनं
वर्गीकृतं प्रष्टुरभीष्टराशिः ॥

i. e. rendered literally into Latin :

*Per multiplicatam radicem diminutæ* [*vel*] *auctæ quantitatis*
*Manifestæ, additæ ad dimidiati multiplicatoris quadratum*
*Radix, dimidiato multiplicatore addito* [*vel*] *subtracto,*
*In quadratum ducta—est interrogantis desiderata*
*quantitas.*

The same is afterwards explained in prose : यो राशिः स्वमूलेन केनचित् गुणितेन ऊनो युतो वा दृष्टस्तस्य मूलस्य गुणार्धकृत्या युक्तस्य दृष्टस्य यत् पदं तद्गुणार्धेन युतं यदि मूलोनो दृष्टो राशिर्भवति यदि गुणघ्नमूलयुतो दृष्टस्तर्हि विहीनं कार्यं तस्य वर्गो राशिः स्यात् ॥ i. e. "A quantity, increased or diminished by its square-root multiplied by some number, is given. Then add the square of half the multiplier of the root to the given number: and extract the square-root of the sum. Add half the multiplier, if the difference were given; or subtract it, if the sum were so. The square of the result will be the quantity sought." (Mr. COLEBROOKE's translation.)

FEIZI's Persian translation of this passage runs thus :

هرگاه شخصي عددي‌را مضمر كرد وجذر اورا يا كسري

( 190 )

از جذر اورا در عددي ضرب كرد ونام مضروب فيه بيان كرد
وحاصل ضرب را با عدد مضمر جمع كرد يا ازوي نقصان كرد
آنچه بعد از جمع يا نقصان حاصل شده است آنرا نيز
ظاهر كرد طريق دانستن آن عدد چنان است كه مضروب
فيه مذكور را تنصيف كرده مجذور او بگيرند وبا حاصل جمع
يا باقي نقصان كه ظاهر كرده بود جمع كرده جذرش بگيرند
بعد از آن نصف مضروب فيه مذكوررا با جذر مذكور جمع
كنند اگر سائل نقصان كرده باشد ونقصان كنند اگر او جمع
كرده است بعد ازآن مجمع يا باقي را مجذور بگيرند بعينه
همان عدد مضمر خواهد بود ۞

With the above Sanskrit stanza from the *Lilavati* some readers will perhaps be interested to compare the following Latin verses, which MONTUCLA (I. p. 590) quotes from LUCAS PACIOLUS:

*Si res et census numero coæquantur, a rebus*
*Dimidio sumpto, censum producere debes,*
*Addereque numero, cujus a radice totiens*
*Tolle semis rerum, census latusque redibit.*

### Page 6, line 16.

فنتّصف الاجذار تكون خمسة ] Such instances of the common instead of the apocopate future, after the imperative, are too frequent in this work, than that they could be ascribed to a mere mistake of the copyist: I have accordingly given them as I found them in the manuscript.

*Page 7, line* 1.

[ وكذلك فانعل ] The same structure occurs page 21, line 15.

*Page 8, line* 11.

[فهذه الستة الضروب] HADJI KHALFA, in his article on Algebra, quotes the following observation from IBN KHALDUN. قال ابن خلدون وقد بلغنا ان بعض ائمة التعاليم من اهل المشرق انتهي المعادلات الي اكثر من هذه الستة وبلغها الي فوق العشرين واستخرج لها كلها اعمالا وثيقة ببراهين هندسية " IBN KHALDUN remarks : A report has reached us, that some great scholars of the east have increased the number of cases beyond six, and have brought them to upwards of twenty, producing their accurate solutions together with geometrical demonstrations."

*Page 8, lime* 17.

See LEONARDO's geometrical illustration of the three cases involving an affected square, as reported by COSSALI (I. p. 2.), and hence by HUTTON (Tracts, &c., II. p. 198.)

CARDAN, in the introduction of his *Ars magna,* distinctly refers to the demonstrations of the three cases given by our author, and distinguishes them from others which are his own. *At etiam demonstrationes, præter tres* MAHOMETIS *et duas* LODOVICI (LEWIS FERRARI, CARDAN's pupil), *omnes nostræ sunt.*—In another passage (page 20) he blames our author for having given the demonstration of only one solution of the case $cx^2 + a = bx$. *Nec admireris,* says he, *hanc secundam demonstrationem aliter quam a* MAHUMETE *explicatam, nam ille immutata figura magis ex re ostendit, sed tamem obscurius, nec nisi unam partem eamque pluribus.*

### Page 17, line 11-13.

The words from وسدس السدس to والا سدسا في are written twice over in the manuscript.

### Page 19, line 12.

[ جذر مال معلوم او اصم ] " The root of a rational or irrational number." In the *Kholáset al Hisáb*, p. 128. 137. 369, the expression منطق (lit. audible) is used instead of معلوم, which stands in a more distinct opposition to اصم (lit. inaudible, surd). BAHA-EDDIN applies the same expressions also to fractions, calling منطق those for which there are peculiar expressions in Arabic, e. g. ثلث one-third, and اصم those which must be expressed periphrastically by means of the word جزء a part, e. g. ثلثة اجزاء من خمسة وعشرين three twenty-fifths. See *Kholáset al Hisáb*, p. 150.

### Page 19, line 15.

The manuscript has مثلي ذلك المال. The context requires the insertion of جذر after مثلي, which I have added from conjecture.

### Page 20, line 15. 17.

[ ما يصيب الواحد ] " What is proportionate to the unit,"

*i. e.* the quotient. This expression will be explained by BAHA-EDDIN's definition of division (*Kholáset al Hisáb*, p. 105). القسمة طلب عدد نسبته الي الواحد كنسبة المقسوم الي المقسوم عليه " Division is the finding a number which bears the same proportion to the unit, as the dividend bears to the divisor."

Page 21, *line* 17.

جذري ] The MS. has جذر .

Page 24, *line* 6.

تمكننا لها صورة لا تحسن ] An attempt at constructing a figure to illustrate the case of $[100+x^2-20x]+[50+10x-2x^2]$ has been made on the margin of the manuscript.

Page 30, *line* 10.

فخذ ما شئت ] A marginal note in the manuscript defines this in the following manner. يعني اقسم العشرة كيف شئت اربعة حنطة وستة شعيرا او ستة حنطة واربعة شعيرا او ثلثة حنطة وسبعة شعيرا اوعكس ذلك او كيف ما شئت فانه يصح العمل فيه حاشية من شرح المزيحفي " He means to say: divide the ten in any manner you like, taking four of wheat and six of barley, or four of barley and six of wheat, or three of wheat and seven of barley, or *vice versa*, or in any other way: for the solution will hold good in all these cases. (*Note from Al Mozaihafi's Commentary*)."

Page 42, *line* 8.

The manuscript has a marginal note to this passage,

from which it appears that the inconvenience attending the solution of this problem has already been felt by Arabic readers of the work.

### Page 45, line 16.

This instance from Mohammed's work is quoted by Cardan (*Ars Magna*, p. 22, edit. Basil.) As the passage is of some interest in ascertaining the identity of the present work with that considered as Mohammed's production by the early propagators of Algebra in Europe, I will here insert part of it. *Nunc autem*, says Cardan, *subjungemus aliquas quæstiones, duas ex Mahumete, reliquas nostras.* Then follows *Quæstio I. Est numerus a cujus quadrato si abjeceris $\frac{1}{3}$ et $\frac{1}{4}$ ipsius quadrati, atque insuper 4, residuum autem in se duxeris, fiet productum æquale quadrato illius numeri et etiam* 12. *Pones itaque quadratum numeri incogniti quem quæris esse* 1 *rem, abjice $\frac{1}{3}$ et $\frac{1}{4}$ ejus, es insuper* 4, *fiet $\frac{5}{12}$ rei* m : 4, *duc in se, fit $\frac{25}{144}$ quadrati* p : 16 m : $3\frac{1}{3}$ *rebus, et hoc est æquali uni rei et* 12; *abjice similia, fiet* 1 *res æqualis $\frac{25}{144}$ quadrati* p : 4 m : $3\frac{1}{3}$ *rebus*, &c.

The problem of the *Quæstio II.* is in the following terms, *Fuerunt duo duces quorum unusquisque divisit militibus suis aureos* 48. *Porro unus ex his habuit milites duos plus altero, et illi qui milites habuit duos minus contigit ut aureos quatuor plus singulis militibus daret; quæritur quot unicuique milites fuerint.* In the present copy of Mohammed's algebra, no such instance occurs. Yet Car-

DAN distinctly intimates that he derived it from our author, by introducing the problem which immediately follows it, with the words: *Nunc autem proponamus quæstiones nostras.*

### Page 46, line 18.

The manuscript has the following marginal note to this passage: هذه المسئلة تعمل بالكعب و طريقه ان تاخذ مالا و تلقي ثلثه يبقي ثلثا مال تضرب ذلك في ثلثة اجذار فيكون كعبين يعدلان مالا فزده مرتين علي قدر المال يكون جذرين يعدلان درهما والجذر نصف المال والمال ربع اذا القيت ثلثه بقي سدس اذا ضربت ذلك في ثلثة اجذاره وهي درهم و نصف بلغ ذلك ربع درهم مثل المال كما ذكر "This instance may also be solved by means of a cube. The computation then is, that you take the square, and remove one-third from it; there remain two-thirds of a square. Multiply this by three roots; you find two cubes equal to one square. Extracting twice the square-root of this, it will be two roots equal to a dirhem. Accordingly one root is one-half, and the square one-fourth.* If you remove one-third of this, there remains one-sixth, and if you multiply this by three roots, that is by one dirhem and a half, it amounts to one-fourth of a dirhem, which is the square as he had stated."

---

\* $[x^2 - \frac{1}{3}x^2] \times 3x = x^2$
$2x^3 = x^2$
$2x = 1$
$x = \frac{1}{2}.$

( 196 )

*Page* 50, *line* 2.

I am uncertain whether my translation of the definition which MOHAMMED gives of mensuration be correct. Though the diacritical points are partly wanting in the manuscript, there can, I believe, be no doubt as to the reading of the passage.

*Page* 51, *line* 12.

I have simply translated the words اهل الهندسة by "geometricians," though from the manner in which MOHAMMED here uses that expression it would appear that he took it in a more specific sense.

FIRUZABADI (Kamus, p. 814, ed. Calcutt.) says that the word *handasah* (الهندسة) is originally Persian, and that it signifies "the determining by measurement where canals for water shall be dug."

The Persians themselves assign yet another meaning to the word هندسه *hindisah*, as they pronounce it: they use it in the sense of decimal notation of numerals.*

It is a fact well known, and admitted by the Arabs

---

* هندسه بكسر اول و ثالث و فتح سين بي نقطه بمعني
اندازه و شكل باشد و ارقامي را نيز گويند كه در زير حروف
كلمات نويسند همچو ابجد هو ز حطي
١.٩٨ ٧٦٥ ۴۳۲۱

"*Hindisah* is used in the sense of measurement and size; the same word is also applied to the signs which are written instead of the words (for numbers) as 1, 2, 3, 4, 5, 6, 7, 8, 9, 10." *Burhani Kati.*

themselves, that the decimal notation is a discovery for which they are indebted to the Hindus.* At what time the communication took place, has, I believe, never yet been ascertained. But it seems natural to suppose that it was at the same period, when, after the accession of the Abbaside dynasty to the caliphat, a most lively interest for mathematical and astronomical science first arose among the Arabs. Not only the most important foreign works on these sciences were then translated into Arabic, but learned foreigners even lived at the court of Bagdad, and held conspicuous situations in those scientific establishments which the noble ardour of the caliphs had called forth. History has transmitted to us the names of several distinguished scholars, neither Arabs by birth nor Mohammedans by their profession, who were thus attached to the court of ALMANSUR and ALMAMUN; and we know from

---

\* It is almost unncessary to adduce further evidence in support of this remark. BAHA-EDDIN, after a few preliminary remarks on numbers, says وقد وضع لها حكماء الهند الارقام التسعة المشهورة "Learned Hindus have invented the well known nine figures for them." (*Kholáset al-Hisáb*, p. 16.) In a treatise on arithmetic, entitled متن النزهة في علم الحساب which forms part of Sir W. OUSELEY's most valuable collection of Oriental manuscripts, the nine figures are simply called الاشكال الهندية. See, on the subject generally, Professor VON BOHLEN's work, *Das alte Indien*, (Königsberg, 1830. 1831. 8.) vol. II. p. 224, and ALEXANDER VON HUMBOLDT's most interesting dissertation: *Ueber die bei verschiedenen Völkern üblichen Systeme von Zahlzeichen*, &c. (Berlin, 1829. 4.) page 24.

good authority, that Hindu mathematicians and astronomers were among their number.

If we presume that the Arabic word *handasah* might, as the Persian *hindisah*, be taken in the sense of decimal notation, the passage now before us will appear in an entirely new light. The اهل الهندسة, to whom our author ascribes two particular formulas for finding the circumference of a circle from its diameter, will then appear to be the Hindu Mathematicians who had brought the decimal notation with them;—and the اهل النجوم منهم, to whom the second and most accurate of these methods is attributed, will be the Astronomers among these Hindu Mathematicians.

This conjecture is singularly supported by the curious fact, that the two methods here ascribed by Mohammed to the اهل الهندسة actually do occur in ancient Sanskrit mathematical works. The first formula, $p = \sqrt{10d^2}$, occurs in the *Vijaganita* (COLEBROOKE's translation, p. 308, 309.); the second, $p = \frac{d \times 62832}{20000}$, is reducible to $\frac{d \times 3927}{1250}$, the proportion given in the following stanza of BHASKARA's *Lilavati*:

व्यासे भनन्दाग्निहते विभक्ते
खवाणसूर्यैः परिधिस्तु सूक्ष्मः ॥ १ ॥
द्वाविंशतिघ्ने विहृते च शैलैः
स्थूलोऽथ वा स्याद्व्यवहारयोग्यः ॥

" When the diameter of a circle is multiplied by three

thousand nine hundred and twenty-seven, and divided by twelve hundred and fifty, the quotient is the near circumference: or multiplied by twenty-two and divided by seven, it is the gross circumference adapted to practice."* (COLEBROOKE's translation, page 87. See FEIZI's Persian translation, p. 126, 127.)

The coincidence of $\frac{d \times 62832}{20000}$ with $\frac{d \times 3927}{1250}$ is so striking, and the formula is at the same time so accurate, that it seems extremely improbable that the Arabs should by mere accident have discovered the same proportion as the Hindus: particularly if we bear in mind, that the Arabs themselves do not seem to have troubled themselves much about finding an exact method.†

---

* The Sanskrit original of this passage affords an instance of the figurative method of the Hindus of expressing numbers by the names of objects of which a certain number is known: the expressions for the units and the lower ranks of numbers always preceding those for the higher ones. भ (lunar mansion) stands for 27; नन्द (treasure of Kuvera) for 9; and अग्नि (sacred fire) for 3: therefore भनन्दाग्नि = 3927. Again, ख (cypher) is 0; वाण (arrow of Kamadeva) stands for 5; सूर्य (the sun in the several months of the year), for 12: therefore खवाणसूर्य = 1250. For further examples, see *As. Res.* vol. XII. p. 281, ed. Calc., and the title-pages or conclusions of several of the Sanskrit works printed at Calcutta;—e. g. the *Sutras* of *Panini* and the *Siddhantakaumudi*.

† This would appear from the very manner in which our author introduces the several methods; but still more from the following marginal note of the manuscript to the present passage: وهو تقريب

*Page* 57, *line* 5-8.

The words between brackets are not in the manuscript: I have supplied the apparent hiatus from conjecture.

*Page* 61, *line* 4.

A triangle of the same proportion is used to illustrate this case in the *Lilavati* (FEIZI's Persian transl. p. 121. COLEBROOKE's transl. of the *Lilavati*, p. 71. and of the *Vijaganita*, p. 203.)

*Page* 65, *line* 12-14.

The words between brackets are in the manuscript written on the margin. I think that the context warrants me sufficiently for having received them into the text.

*Page* 66, *line* 5.

The words between brackets are not in the text, I give them merely from my own conjecture.

---

لا تحقيق ولا يقف احد علي حقيقة ذلك ولا يعلم دورها الا الله لان الخط ليس بمستقيم فيوقف علي حقيقته وانما قيل ذلك تقريب كما قيل في جذر الاصم انه تقريب لا تحقيق لان جذره لا يعلمه الا الله واحسن ما في هذه الاقوال ان تضرب القطر في ثلثة وسبع لانه اخف واسرع والله اعلم ❊ " This is an approximation, not the exact truth itself: nobody can ascertain the exact truth of this, and find the real circumference, except the Omniscient: for the line is not straight so that its exact length might be found. This is called an approximation, in the same manner as it is said of the square-roots of irrational numbers that they are an approximation, and not the exact truth: for God alone knows what the exact root is. The best method here given is, that you multiply the diameter by three and one-seventh: for it is the easiest and quickest. God knows best!"

( 201 )

## Page 71, line 8, 9.

The author says, that the capital must be divided into 219320 parts: this I considered faulty, and altered it in my translation into 964080, to make it agree with the computation furnished in the note. But having recently had an opportunity of re-examining the Oxford manuscript, I perceive from the copious marginal notes appended to this passage, that even among the Arabian readers considerable variety of opinion must have existed as to the common denominator, by means of which the several shares of the capital in this case may be expressed.

One says: انظر لمال يكون لسدسه ربع والربعه ثلث وما بقي يتقسم علي ماية و خمسة و تسعين ولا يوجد ذلك في اقل من اربعة وعشرين فاضرب اربعة وعشرين في ماية وخمسة و تسعين يصح من ذلك اربعة الاف وستماية وثمانون ومنه يصح
" Find a number, one-sixth of which may be divided into fourths, and one-fourth of which may be divided into thirds; and what thus comes forth let be divisible by hundred and ninety-five. This you cannot accomplish with any number less than twenty-four. Multiply twenty-four by one hundred and ninety-five: you obtain four thousand six hundred and eighty, and this will answer the purpose."

Another:* وفي وجه اخر انك تجعل ماية وستة وخمسين

---

\* The numbers in this and in part of the following scholium are in the MS. expressed by figures, which are never used in the text of the work.

2 D

سدس المال وتضربها في ٦ فيكون ٩٣٦ واذا استخرجت نصيب الابن وهو الثلث والربع وجدته ٥٤٦ و لا خمس لها فاضربها في ٥ يكون ٤٦٨٠ للام من ذلك ٤٢٥ وللزوج ٧٨٠ وللابن ٢٨٨ ولصاحب الخمسين ١٤٩٢ والصاحب الربع ٦٩٥ "Ac-cording to another method, you may take one hundred and fifty-six for the one-sixth of the capital. Multiply this by six; you find nine hundred and thirty-six. Taking from this the share of the son, which is one-third and one-fourth, you find it five hundred and forty-six. This is not divisible by five: therefore multiply the whole number of parts by five: it will then be four thousand six hundred and eighty. Of this the mother receives four hundred and twenty-five, the husband seven hundred and eighty, the son two hundred and eighty-eight (twelve hundred and eighty-eight?), the legatee, who is to receive the two-fifths, fourteen hundred and ninety-two, and the legatee to whom the one-fourth is bequeathed, six hundred and ninety-five."

Another: وفي [وجه] اخر يصح من تسعة الاف و ثلثماية وستين ووجه العمل في ذلك ان [تقسم] الفريضة في اثني عشر للام سهمان وللزوج ثلثة وللابن سبعة فتضربها في ٢ لذكر الخمسي والربع فيكون مايتين واربعين فتاخذها سدسها اربعين للام والثلث جائز عليها وليس للاربعين ثلث فتضرب اصل المسئلة في ثلثة لذلك فيكون سبعماية وعشرين فتاخذ سدسها للام ماية وعشرين فيخرج من ذلك الثلث لاصحاب الوصايا وهو اربعون مقسوم علي ثلثة عشر لا يصح فاضرب المسئلة في

[MS. ٩٠٦٣] ٩٣٦٠ يكون ١٣ لما ذكرنا للام من ذلك ثماني ماية و خمسون و للابن الفان و خمسماية و ستة و سبعون و للزوج الف و خمسماية و ستون ولصاحب الخمسين الفان و تسعماية واربعة وثمانون ولصاحب الربع الف و ثلثماية و تسعون والله اعلم " According to another method, the number of parts is nine thousand three hundred and sixty. The computation then is, that you divide the property left into twelve shares; of these the mother receives two, the husband three, and the son seven. This (number of parts) you multiply by twenty, since two-fifths and one-fourth are required by the statement. Thus you find two hundred and forty. Take the sixth of this, namely forty, for the mother. One-third out of this she must give up. Now, forty is not divisible by three. You accordingly multiply the whole number of parts by three, which makes them seven hundred and twenty. The one-sixth of this for the mother is one hundred and twenty. One-third of this, namely forty, goes to the legatees, and should be divided by thirteen; but as this is impossible, you multiply the whole number of parts by thirteen, which makes them nine thousand three hundred and sixty, as we said above. Of this the mother receives eight hundred and fifty, the son two thousand five hundred and seventy-six, the husband one thousand five hundred and sixty, the legatee to whom the two-fifths are bequeathed, two thousand nine hundred and eighty-four, and the legatee who is to receive one-fourth, one thousand three hundred and ninety."

Another scholium briefly says: مِن مايةِ الفٍ واحدٍ وعشرين الفٍ وستمايةٍ وثمانينَ في لفظِ شرحِ المزيحفي فاذا اردتَ اختصارها فارجعها الي نصفِ جزءٍ من ثلثةِ عشر "With one hundred and twenty-one thousand six hundred and eighty, according to MOZAIHAFI's commentary. If you want to express it briefly, you may reduce it by taking moieties of thirteenths."

*Page* 85, *line* 8.

The manuscript has the following marginal note to this passage: وانْ شئتَ في عملِ هذهِ المسئلةِ فاجعلِ الوصيةَ الاولي نصيبا لانه اوصي له بنصيبٍ ولم يستثنِ عليه شيئًا واجعلِ الوصيتينِ الاخرينِ شيئًا وزدْ ذلكَ علي انصباءِ الورثةِ يكونُ الجميعُ سبعةَ انصباءٍ وشيئًا واعملْ علي ما تقدمَ تخرجْ النصيبُ ۴۹ والشيءُ ۵۳ "If you prefer, you may also, in solving this problem, make the first legacy a share, since the testator has bequeathed a whole share without any deduction; and call the two other legacies thing. Add this to the shares of the heirs: the total amount will be seven shares and thing. Then proceed as above: you will find the share to be forty-nine, and the thing fifty-three."

*Page* 93, *line* 1.

The following is a marginal note of the manuscript: معني السؤالِ في هذهِ التكملةِ انَّ قولَه بتكملةِ خمسِ المالِ بنصيبِ بنتٍ اي اوصتْ له بخمسِ المالِ الا نصيبَ بنتٍ وذلكَ قوله بتكملةِ ربعِ المالِ بنصيبِ الامِ اي اوصتْ له بربعِ المالِ الا نصيبَ الامِ "The purpose of the question about such a completement is this. If the author says: *as much as must be added to the share of a daughter to make it equal to one-fifth of the capital*, he means to say, that the testatrix bequeathed to the legatee one-fifth of the capital, less the share of the daughter; again, if he says: *as much as must be added to the share of the mother to make it one-fourth of the capital*, he intends, that the testatrix bequeathed to the legatee one-fourth of the capital, less the share of the mother."

*Page 95, line 14, 19.*

The manuscript has here the following note. قال الفقيه
احمد بن عباس (*) اقول ان التكملة في هذه المسئلة ١٣ سهما والاستثناء من التكملة هو ربع ما يبقي من المال بعد رفع التكملة من المال والذي يبقي من المال بعد رفع التكملة منه ٥٦ وربعها ١٤ اذا نزعت منها نصيب بنت وهو ٥ بقي منها ٩ وهي الاستثناء من التكملة اذا رفعتها من التكملة وهي ١٣ "  بقي منها اربعة وهو الوصية كما ذكر والله اعلم  The Fakih AHMED BEN ABBAS (*) says: I hold, that the completement in this instance is thirteen parts, and the deduction from the completement is one-fourth of what remains of the capital after the completement has been taken from the capital. This remainder of the capital, after subtracting the completement, is fifty-six, and its fourth is fourteen. If you subtract from this the portion of a daughter, which is five,

---

\* The name is written very indistinctly in the manuscript.

there remains nine of it, and this is the deduction from the completement. Subtracting it from the completement, which is thirteen, there remains four, and this is the legacy, as the author has said."

*Page 98, line 8.*

The word مثلها which I have omitted in my translation of this and of two following passages, is in the manuscript explained by the following scholium : مثلها متساوية لها في الحسن و السن و النسب والمال والبلد والعصر..... والبكارة " Adequate, *i. e.* corresponding to her beauty, her age, her family, her fortune, her country, the state of the times, .... and her virginity." (Part of the gloss is to me illegible.) The dowry varies according to any difference in all the circumstances referred to by the scholium. See HAMILTON's Hedaya, vol. I. page 148.

*Page 113, line 7.*

The manuscript has the following marginal note (?). العقر في الامة بمنزلة مهر المثل في الحرة وهو ما تتزوج عليه مثلها في الاوصاف المعتبرة في المماثلة " The *Okr* of a slave girl corresponds to the adequate dowry of a free-born woman; it is a sum of money on payment of which one of distinguished qualities corresponding to her would be married." See HAMILTON's Hedaya, vol. II. page 71.

I am very doubtful whether I have well understood the words in which our author quotes ABU HANIFAH's opinion.

ABU HANIFAH AL NO'MAN BEN THABET is well known

as an old Mohammedan lawyer of high authority. He was born at Kufa, A.H. 80 (A.D. 690), and died A.H. 150 (A.D. 767). EBN KHALLIKAN has given a full account of his life, and relates some interesting anecdotes of him which bear testimony to the integrity and independence of his character.

*Page* 113, *line* 16.

The marginal notes on this chapter of the manuscript give an account of what the computation of the cases here related would be according to the precepts of different Arabian lawyers, e. g. SHAFEI, ABU YUSSUF, &c. The following extract of a note on the second case will be sufficient as a specimen: الجواب الذي ذكره الخوارزمي في هذه المسئلة انما هو علي مذهب ابي يوسف وزفر (*) واحد الوجوه لاصحاب الشافعي فاما ابو حنيفة فانه يجعل ما لزم الواهب بسبب وطئه وصية ايضا فتكون الوصية علي قوله شيئا و ثلثا وهو احد الوجوه علي مذهب الشافعي وعند محمد بن الحبيس (*) تجعل وطء الواهب لما وهب منه و الا يلزمه شيء بسبب ذلك وهو احد الوجوه علي هذهب الشافعي فعلي هذا الوجه تصح الهبة في ثلثها و تبطل في ثلثيها ولا دور لان التركة علي حالها وعلي قول ابن حنيفة تعمل لما فعلت علي مذهب ابي يوسف وزفر (*)فاذا صار بايدي الورثة ثلثماية الا شيئا و ثلث شيء يعدل شيئين و ثلثي شيء لان الذي لزمه بالعقر وصية ايضا فاذا جبرت وقابلت عدل الشيء خمسة وسبعين درهما وهو ربع للجارية فتصح الهبة في ربعها وتبطل

---

\* These names are very indistinctly written in the manuscript.

في ثلثة ارباعها " The solution of this question given by the Khowarezmian is according to the school of ABU YUSSUF WAZFAR, and one of the methods of SHAFEI's followers. ABU HANIFAH calls the sum which the donor has to pay on account of having cohabited with the slave-girl likewise a legacy; thus, according to him, the legacy is one and one-third of thing: this is another method of SHAFEI's school. According to MOHAMMED BEN AL JAISH, the donor has nothing to pay on account of having cohabited with the slave girl:* and this is again a method adopted by the school of SHAFEI. After this method, one-third of the donation is really paid, whilst two-thirds become extinct: and there is no return, as the heritage has remained unchanged. According to ABU HANIFAH, you proceed in the same manner as after the precepts of ABU YUSSUF WAZFAR. Thus the heirs obtain three hundred less one and one-third of thing, which is equal to two things and two-thirds: for what he (the donor) has to pay on behalf of the dowry, is likewise a legacy. Completing and reducing this, one thing is equal to seventy-five dirhems: this is one-fourth for the slave-girl; one-fourth of the donation is actually paid, and three-fourths become extinct."

* I doubt whether this is the meaning of the original, the words from محمد till يلزمه being very indistinctly written in the MS.

## غلط نامه

| صحیح | غلط | سطر | صفحه |
|---|---|---|---|
| والمال والمالین | والمالین | ۱۲ | ۲۴ |
| وتخف | وتحق | ۶ | ۲۵ |
| في الآخر | والآخر | ۱۴ | |
| وعشرین | وعشرة | ۱۶ | ۳۱ |
| شعیراً | شعیر | ۸ | ۳۵ |
| تنصیف | تصنیف | ۸ | ۴۱ |
| مثلي | مثل | ۹ | ۴۲ |
| خمساه وربعه | خمسان وربعة | ۱۵ | ۶۵ |
| وبثلثِ | وبثلثي | ۱۵ | ۷۲ |
| وثلث | وثلثي | ۱ | ۷۳ |
| أن تقیم | تقیم | ۱۹ | — |
| من ثلثین جزءًا من سهم فزد | من سهم فزده | ۱۵ | ۷۵ |
| خمسي | خمس | ۱۳ | ۸۱ |
| اربعة | الانصبا اربعة | ۴ | ۸۷ |
| وثلثة | وثلثي | ۳ | ۹۰ |
| وهو | هو | ۱۲ | — |
| ثلثة | ثلثه | ۷ | ۹۱ |
| من مایتین واربعین سهما من مال | من مال | ۱۱ | ۹۲ |
| فتجد | فخذ | ۱۷ | ۹۴ |
| بثلثي | فثلثي | ۱۶ | ۹۹ |
| وصیتها | وصیتک | ۴ | ۱۰۰ |
| الا شیئًا | الا شيء | ۹ | — |
| ونصفاً | ونصف | ۱۱ | — |
| عبدًا | عبد | ۷ | ۱۰۲ |
| مثلي | مثلاً | ۱۶ | ۱۰۸ |
| مایتاً | مایتی | ۱۱ | ۱۱۱ |
| وثلث | وثلثا | ۱۳ | ۱۱۲ |
| فالشيء | وشيء | ۱۴ | ۱۱۶ |

درهما وشيٌ ونصف شيٍ فمثل نصفها هو الوصية وهو عشرة دراهم وثلثة ارباع شيٍ وذلك ثلث المال وهو ستة عشر درهما وثلثا درهم فالق عشرة بعشرة فيبقى ستة دراهم وثلثان يعدل ثلثة ارباع شيٍ فكمّل الشيَ وهو ان تزيد عليه ثلثه وزد علي الستة والثلثين ثلثها وهو درهمان وتسعا درهم فيكون ثمانية دراهم وثمانية اتساع درهم يعدل شيئا فانظر كم الثمانية الدراهم والثمانية الاتساع من راس المال وهو عشرون درهما فتجد ذلك اربعة اتساعها فرد من الكر اربعة اتساعه وترك خمسة اتساع العشرين فيكون قيمة اربعة اتساع الكر اثني وعشرين درهما وتسعي درهم وخمسة اتساع العشرين احد عشر درهما وتسع درهم فيصير في ايدي الورثة ثلثة وثلثون درهما وثلث درهم وهو ثلثا الخمسين الدرهم * والله اعلم *
تم الكتاب بحمد الله ومنه وتوفيقه وتشديده *

## باب السلم في المرض *

اذا اسلم رجل في مرضه ثلثين درهما في كرّ من طعام يساوي عشرة دراهم ثم مات في مرضه فانه يرد الكرّ ويرد علي ورثة الميت عشرة دراهم قياسه ان يرد الكرّ و قيمته عشرة دراهم فيكون قد حاباه بعشرين درهما فالوصية من المحاباة شيء و يصير في ايدي الورثة عشرون غير شيء و كرّ وكل ذلك ثلثون درهما غير شيء يعدل شيئين وهو مثلا الوصية فاجبر الثلثين بالشيء وزده علي الشيئين فيصير الثلثون يعدل ثلثة اشياء الشيء من ذلك ثلثه وهو عشرة دراهم وهو ما جاز من المحاباة *

فان اسلم الي رجل عشرين درهما وهو مريض في كرّ يساوي خمسين درهما ثم اقاله في مرضه ثم مات فانه يرد اربعة اتساع الكرّ وأحد عشر درهما وتسع درهم وقياسه انك قد علمت ان قيمة الكرّ مثل الذي اسلم اليه مرتين ونصفا فهو لا يرد من راس المال شيئًا الا ردّ من الكرّ مثليه ومثل نصفه فتجعل الذي يرد من الكرّ بالشيء فشيئين فنصفا فزده علي ما بقي من العشرين وهو عشرون غير شيء فيصير في ايدي ورثة الميت عشرون كرّ

فيكون بعض الشيء و ثلثين درهما يعدل نصف شيء فيكون نصف شيء غير ثلثين يعدل بعض الشيء الذي هو وصية الموهوب له للواهب فاعرف ذلك ثم ارجع الي ما بقي في يد الواهب وهو ثلثماية غير شيء وصار اليه بعض الشيء وهو نصف الشيء الا ثلثين درهما فيبقي في يده مايتان وسبعون غير نصف شيء واخذ العقر وهو ماية درهم غير ثلث شيء ورد العقر وهو ثلث ما بقي من الشيء بعد رفع بعض الشيء منه وهو سدس شيء وعشرة دراهم فحصل في يده ثلثماية وستون غير شيء وذلك مثلا الشيء والعقر الذي رد فنصف ذلك ماية وثمانون غير نصف شيء وهو مثل الشيء والعقر فاجبر ذلك بنصف شيء وزده علي الشيء والعقر فيكون ماية وثمانين درهما يعدل شيئا و نصف شيء والعقر الذي رد وهو سدس شيء وعشرة دراهم تسقط عشرة فيبقي ماية وسبعون درهما يعدل شيئا وثلثي شيء فاردده لتعرف الشيء وهو ان تاخذ ثلثة اخماسه فيكون ماية واثنين يعدل الشيء الذي هو وصية الواهب للموهوب له واما وصية الموهوب له للواهب فهو نصف ذلك غير ثلثين درهما وهو احد وعشرون والله اعلم *

شیئان و ثلثي شيء فاجبر ذلك بثلثة اشیاء فیكون اربعماية يعدل ثمانية اشياء وثلث شيء فقابل بذلك فيكون الشيء الواحد يعدل ثمانية واربعين درهما  *

فان قال رجل وهب لرجل جارية في مرضه تيمتها ثلثماية درهم وعقرها ماية درهم فوطئها الموهوب له ثم وهبها الموهوب له للواهب في مرضه ايضا فوطئها الواهب كم جاز منها وكم انتقص فقياسه ان تجعل تيمتها ثلثماية درهم و الوصية من ذلك شيء فيبقي في ايدي ورثة الواهب ثلثماية غير شيء و صار في يد الموهوب له شيء واعطا الموهوب له الواهب بعض الشيء و بقي في يده غير بعض شيء ورث اليه ماية غير ثلث شيء واخذ العقر ثلث شيء غير ثلث بعض شيء فصار في يده شيء و ثلثا شيء غير ماية درهم وغير بعض شيء وغير ثلث بعض الشيء و ذلك مثلا بعض الشيء فنصفه مثل بعض الشيء وهو خمسة اسداس شيء غير خمسين درهما و غير ثلثي بعض شيء فاجبر ذلك بثلثي بعض الشيء وبخمسين درهما فيكون خمسة اسداس شيء تعدل بعض شيء و ثلثي بعض شيء وخمسين درهما فاردد ذلك الي بعض شيء لتعرفه وهو ان تاخذ ثلثة اخماس

بثلث ماله فان قول ابي حنيفة الثلث بينهما نصفان
و قياسه ان تجعل الوصية للموهوب له الجارية شيئا فيبقي
ثلثماية غير شيء ثم رد العقر وهو ثلث شيء فيبقي معه
ثلثماية غير شيء و ثلث شيء فوصيتة في قول ابي
حنيفة شيء وثلث شيء و في قول الاخر شيء ثم تعطي
الموصي له بالثلث مثل وصية الاول وهو شيء وثلث شيء
فيبقي في يده ثلثماية غير شيئين و ثلثي شيء يعدل
مثلي الوصيتين وهما شيئان و ثلثا شيء فنصف ذلك
يعدل الوصيتين وهو ماية . و خمسون غير شيء وثلث
شيء فاجبر ذلك بشيء و ثلث شيء وزده علي
الوصيتين فصار ماية و خمسين يعدل اربعة اشياء فالشيء
من ذلك ربعه وهو سبعة و ثلثون و نصفا *

فان قال و طئها الموهوب له و وطئها الواهب واوصي
بثلث ماله * فان القياس في قول ابي حنيفة ان
تجعل الوصية شيئا فيبقي ثلثماية غير شيء واخذ العقر
ماية غير ثلث شيء فصار في يده اربعماية درهم غير
شيء و ثلث شيء ورد العقر ثلث شيء واعطا الموصي
له بالثلث مثل وصية الاول شيئا وثلث شيء فيبقي
اربعماية درهم غير ثلثة اشياء يعدل مثلي الوصية وذلك

فصار في ايدي ورثة الواهب ثلثماية غير شيء و ثلث شيء وذلك مثلا الوصية التي هي شيء وهو شيئان فاجبر ذلك بشيء و ثلث شيء وزده علي الشيئين فيكون ثلثماية يعدل ثلثة اشياء وثلث شيء فالشيء من ذلك ثلثة اعشاره وهو تسعون درهما و ذلك الوصية *

فان كانت المسئلة علي حالها و وطئها الواهب والموهوب له فقياسه ان تجعل الوصية شيئا والمنتقص ثلثماية غير شيء و يلزم الواهب للموهوب له العقر بالوطيء ثلث شيء و يلزم الموهوب له ثلث الانتقاص وهو ماية غير ثلث شيء فصار في ايدي ورثة الواهب اربعماية غير شيء وثلثي شيء وذلك مثلا الوصية فاجبر الاربعماية بشيء وثلثي شيء وزدها علي الشيئين فيكون اربعماية يعدل ثلثة اشياء وثلثي شيء فالشيء من ذلك ثلثة اجزاء من احد عشر جزءا من اربعماية وهو ماية وتسعة وجزؤ من احد عشر من درهم وذلك الوصية والانتقاص ماية و تسعون وعشرة اجزاء من احد عشر جزءا من درهم * وفي قول ابي حنيفة تجعل الشيء وصية وما صار اليه بالعقر ايضا وصية *

فان كانت المسئلة علي حالها فوطئها الواهب واوصي

درهم يعدل شيئين وتسعة وعشرين جزءًا من اربعين جزءًا من شيء فقابل به فيكون الشيء يعدل ثلثة وسبعين درهما وثلثة واربعين جزءًا من ماية وتسعة اجزاء من درهم *

باب العقر في الدور *

رجل وهب لرجل جارية في مرض موته ولا مال له غيرها ثم مات وقيمتها ثلثماية درهم وعقرها ماية درهم فوطئها الرجل الموهوب له فقياسه ان تجعل الوصية الموهوب له الجارية شيئًا فتنقص من الهبة ثلثماية غير شيء ويرجع الي ورثة الواهب ثلث الانتقاص للعقر لان العقر ثلث القيمة وذلك ماية درهم غير ثلث شيء فصار في ايدي ورثة الواهب اربعماية غير شيء وثلث شيء وذلك مثلا الوصية التي هي شيء وذلك شيئان فاجبر الاربعماية بشيء وثلث شيء وزده علي الشيئين فيكون اربعماية يعدل ثلثة اشياء وثلث شيء وشيء من ذلك ثلثة اعشاره وهو ماية وعشرون درهما وهي الوصية *

فان قال وهبها في مرضه وقيمتها ثلثماية وعقرها ماية فوطئها الواهب ثم مات فقياسه ان تجعل الوصية شيئًا والمنتقص ثلثماية غير شيء فوطئها الواهب فلزمه العقر وهو ثلث الوصية لان العقر ثلث القيمة وهو ثلث شيء

عشرة اجزاء من واحد وثلثين جزءا من درهم فالوصية من المايتين على قدر ذلك وهي اربعة وستون درهما وستة عشر جزءا من واحد وثلثين جزءا من الدرهم *

فان اعتق جارية قيمتها ماية درهم و وهب لرجل جارية قيمتها خمسماية درهم فوطئها الموهوب له وعقرها ماية درهم واوصى الواهب لرجل بربع ماله فقول ابي حنيفة ان صاحب الجارية لا يضرب باكثر من الثلث و صاحب الربع يضرب بالربع * وقياسه ان قيمة الجارية خمسماية درهم والوصية من ذلك شيء فيبقى خمسماية درهم غير شيء واحد و العقر ماية درهم غير خمس شيء فصار في ايدي الورثة ستماية درهم غير شيء وخمس شيء ثم تعزل وصية صاحب الربع ثلثة ارباع شيء لان الثلث اذا كان شيئا فالربع ثلثة ارباعه فيبقى ستماية درهم غير شيء و ثمانية وثلثين جزءا من اربعين جزءا من شيء وذلك مثلا الوصية فنصف ذلك يعدل وصاياهم وهي ثلثماية درهم غير تسعة وثلثين جزءا من اربعين جزءا من شيء فاجبر ذلك بهذه الاجزاء فيكون ثلثماية درهم يعدل ماية درهم وشيئين وتسعة وعشرين جزءا من اربعين جزءا من شيء فاطرح ماية بماية فيبقى مايتا

فقابل بذلك فتجد الشيء من ذلك خمسة اثمانه فتاخذ خمسة اثمان مايتين وهو ماية و خمسة و عشرون وهو الشيء و ذلك وصية الذي اوصي له بالجارية *

فان اعتق عبدا له قيمته ماية درهم و وهب لرجل جارية قيمتها خمسماية درهم و عقرها ماية درهم فوطئها الموهوب له الواهب و اوصي لرجل بثلث ماله فقياسه في قول ابي حنيفة انه لا يضرب صاحب الجارية باكثر من الثلث فيكون الثلث بينهما نصفين * وقياسه ان تجعل قيمة الجارية خمسماية درهم الوصبة من ذلك شيء فصار في ايدي الورثة من ذلك خمسماية درهم غير شيء واحد و العقر ماية غير خمس شيء فصار في ايديهم ستماية غير شيء و خمس شيء واوصي لرجل بثلث ماله وهو مثل وصية صاحب الجارية وهو شيء فيبقي في ايدي الورثة ستماية غير شيئين و خمس شيء و ذلك مثلا وصاياهم جميعا قيمة العبد والشيئين الموصي بهما فنصف ذلك يعدل وصاياهم وهو ثلثماية غير شيء وعشر شيء فاجبر ذلك بشيء و عشر شيء فيكون ثلثماية يعدل ثلثة اشياء وعشر شيء وماية درهم فاطرح ماية بماية فيبقي مايتان يعدل ثلثة اشياء وعشر شيء فقابل به فالشيء من ذلك

سبعة و عشرين جزءًا من شيء فقابل به و تحطه الي شيء واحد و ذلك ان تنقص منه سبعة اجزاء من اربعة و ثلثين جزءًا منه فيكون الشيء الواحد يعدل مايتي درهم و عشرة دراهم و خمسة اجزاء من سبعة عشر جزءًا من درهم و هو الوصية *

فان اعتق عبدًا له في مرضه قيمته ماية درهم و وهب لرجل جارية قيمتها خمسماية درهم و عقرها ماية درهم فوطئها الموهوب له * فقول ابي حنيفة ان العتق اولي فتبدا به و قياسه ان تجعل قيمة الجارية خمسماية درهم في قوله و قيمة العبد ماية درهم و تجعل وصية صاحب الجارية شيئًا اخر فقد امضي عتق العبد و قيمته ماية درهم و اوصي للموهوب له بشيء و زد العقر ماية درهم غير خمس شيء فصار في ايدي الورثة ستماية درهم غير شيء و خمس شيء و هو مثلا الماية الدرهم و الشيء فنصف ذلك مثل وصيتهما و هو ثلثماية غير ثلثة اخماس شيء فاجبر الثلثماية بثلثة اخماس شيء و زد مثلها علي الشيء فيكون ذلك ثلثماية درهم يعدل شيئًا و ثلثة اخماس شيء و ماية درهم فاطرح من الثلثماية ماية بماية فيبقي مايتا درهم يعدل شيئًا و ثلثة اخماس شي

وستون درهما وثلثان وثلث شيء ولابنته مثل ذلك تضمه الي ما تركت وهو ثلثماية درهم فيكون ثلثماية وستة وستون درهما وثلثي درهم وثلث شيء وقد اوصت بثلث مالها وهو ماية درهم واثنان وعشرون درهما وتسعا درهم وتسع شيء ويبقي مايتان واربعة واربعون واربعة اتساع درهم وتسعا شيء للام من ذلك الثلث واحد وثمانون درهما واربعة اتساع وثلث تسع درهم وثلثا تسع شيء ورجع ما بقي الي السيد وهو ماية واثنان وستون درهما وثمانية اتساع وثلثا تسع درهم وتسع شيء وثلث تسع شيء ميراثا له لانه حصته فحصل في ايدي ورثة السيد خمسماية وتسعة وعشرون درهما وسبعة عشر جزءا من سبعة وعشرين جزءا من درهم غير اربعة اتساع شيء وثلثا تسع شيء وذلك مثلا الوصية التي هي شيء فنصف ذلك مايتان واربعة وستون درهما واثنان وعشرون جزءا من سبعة وعشرين جزءا من درهم غير سبعة اجزاء من سبعة وعشرين جزءا من شيء فاجبر ذلك بالسبعة الاجزاء وتزيد عليها الشيء فيكون ذلك مايتين واربعة وستون درهما واثنين وعشرين جزءا من سبعة وعشرين جزءا من درهم يعدل شيئا وسبعة اجزاء من

السعاية ثلثمائة غير شيء فيبقي شيء للبنت نصفه وللسيد نصفه فتضيف حصة البنت وهي نصف شيء الي تركتها وهي ثلثمائة فيكون ثلثمائة درهم ونصف شيء للزوج من ذلك النصف ويرجع الي السيد النصف وهو مائة وخمسون وربع شيء فصار جميع ما في يد السيد اربعمائة وخمسين غير ربع شيء فذلك مثلا الوصية فنصف ذلك مثل الوصية وهو مايتان وخمسة وعشرون درهما غير ثمن شيء يعدل شيئا فاجبر ذلك بثمن شيء وزده علي الشيء فيكون مايتين وخمسة وعشرين درهما يعدل شيئا وثمن شيء فقابل بذلك فالشيء الواحد ثمانية اتساع مايتين وخمسة وعشرين و ذلك مايتي درهم *

فان اعتق عبدا له في مرضه قيمته ثلثمائة درهم فمات العبد و ترك خمسمائة درهم و ترك بنتا واوصي بثلث ماله ثم ماتت البنت وتركت امها واوصت بثلث مالها وتركت ثلثمائة درهم فقياسه ان ترفع من تركة العبد السعاية وهي ثلثمائة درهم غير شيء فيبقي مايتا درهم و شيء وقد اوصي بثلث ماله وهو ستة وستون درهما وثلثان وثلث شيء ويرجع الي السيد بميراثه ستة

غير ثلث شيء ثم تقضي من ذلك دين المولي وهو ثلثماية درهم فيبقي سبعماية درهم غير ثلث شيء وهو مثلا وصية العبد وهي شيء فنصف ذلك ثلثماية وخمسون غير سدس شيء يعدل شيئا فاجبر ذلك بسدس شيء فيكون ثلثماية وخمسين يعدل شيئا وسدس شيء فيكون الشيء ستة اسباع الثلثماية والخمسين وهو ثلثماية درهم وذلك الوصية فتجمع تركة العبد وما استهلك المولي وهو الفان وثلثماية وخمسون درهما فتعزل من ذلك الدين مايتي درهم ثم تعزل السعاية وهي قيمة الرقبة غير الوصية مايتا درهم فيبقي الف وتسعماية درهم وخمسون درهما للام من ذلك الثلث ستماية درهم و خمسون درهما فالقه والتي الدين وهو مايتا درهم من تركة العبد الموجودة وهي الف وسبعماية و خمسون درهما فيبقي تسعماية درهم تقضي منها دين المولي ثلثماية درهم ويبقي ستماية درهم وذلك مثلا الوصية *

_____

فان اعتق عبدا له في مرضه قيمته ثلثماية درهم ثم مات العبد وترك بنتا وترك ثلثماية درهم ثم ماتت البنت و تركت زوجا و تركت ثلثماية درهم ثم مات السيد فقياسه ان تجعل تركة العبد ثلثماية درهم وتجعل

العبد وما تعجل منه المولي وذلك الف و خمسماية درهم فترفع من ذلك السعاية وهي مايتان و عشرون درهما فيبقي الف ومايتان وثمانون درهما. درهما للابنة النصف ستماية واربعون درهما فتلقيه من تركة العبد وهي الف درهم فيبقي ثلثماية وستون درهما فتقضي من ذلك دين المولي مايتا درهم ويبقي في ايدي الورثة ماية وستون درهما و ذلك مثلا الوصية  *

فان اعتق عبدا له في مرضه قيمته خمسماية درهم فتعجل منه ستماية درهم فاستهلكها و علي المولى دين ثلثماية درهم ثم مات العبد وترك امه ومولاه و ترك الفا و سبعماية و خمسين درهما وعلي العبد دين مايتا درهم فقياسه ان تجعل تركة العبد الفا وسبعماية وخمسين درهما والذي تعجل المولي وهو ستماية درهم فذلك الفان وثلثماية وخمسون درهما فتعزل منه الدين مايتي درهم وتعزل منه السعاية خمسماية درهم غير شيء والوصية شيء فيبقي الف وستماية وخمسون درهما وشيء للام من ذلك الثلث خمسماية و خمسون و ثلث شيء فتلقيه هو والدين الذي هو مايتا درهم من تركة العبد الموجودة وهي الف وسبعماية وخمسون فيبقي الف درهم

ثلثماية وماىتان استهلكها المولي وذلك خمسماية درهم فيعطي المولي السعاية وهي ماىتان وعشرون درهما ويبقي ماىتان وثمانون للابنة النصف من ذلك ماىة واربعون درهما فتلقىه من تركة العبد وهي ثلثماية فيبقي في ايدي الورثة ماىة وستون درهما وذلك مثلا وصية العبد التي هي شيء ٭

فان اعتق عبدا له في مرضه قيمته ثلثماية درهم وقد تعجل المولي منه خمسماية درهم ثم مات العبد قبل موت المولي وترك الف درهم وترك ابنة وعلي المولي دين ماىتا درهم فقياسه ان تجعل تركة العبد الف درهم فالخمسماية التي استهلكها المولي السعاية من ذلك ثلثماية غير شيء فيبقي الف وماىتان وشيء والنصف من ذلك لابنة العبد وهو ستماية درهم ونصف شيء فتلقىه من تركة العبد وهي الف درهم فيبقي اربعماية درهم غير نصف شيء تقضي من ذلك دين المولي وهو ماىتا درهم فيبقي ماىتا درهم غير نصف شيء يعدل مثلا الوصية التي هي الشيء وذلك شيئان فاجبر ذلك بنصف شيء فيكون ماىتي درهم يعدل سىئين ونصفا فقابل به فالشيء يعدل ثمانين درهما وهي الوصية فتجمع تركة

و نصف شيء فيصير سبعماية درهم يعدل خمسة اشياء و نصف شيء فقابل به فيصير الشيء الواحد ماية وسبعة و عشرين درهما و ثلثة اجزاء من احد عشر من درهم *

فان اعتق عبدا له في مرضه قيمته ثلثماية درهم وقد تعجل المولي منه مايتي درهم فاستهلكها ثم مات العبد قبل موت السيد و ترك بنتا و ترك ثلثماية درهم فقياسه ان تجعل تركة العبد الثلثماية و المايتين اللتين استهلكهما المولي فذلك خمسماية درهم فتعزل منها السعاية وهي ثلثماية غير شيء لان وصيته شيء فيبقي مايتا درهم و شيء للابنة من ذلك النصف ماية درهم و نصف شيء و يرجع الي ورثة السيد النصف بالميراث وهو ماية درهم و نصف شيء في ايديهم من الثلثماية و الدرهم غير شيء ماية درهم غير شيء لان المايتين مستهلكتان فيبقي في ايديهم بعد المايتين المستهلكين مايتا درهم غير نصف شيء و ذلك يعدل وصية العبد مرتين فنصفها ماية غير ربع شيء يعدل وصية العبد وهي شيء فتجبر ذلك بربع شيء فيكون ماية درهم يعدل شيئا و ربع شيء فالشيء من ذلك اربعة اخماس وهو ثمانون درهما وهي الوصية و السعاية مايتان و عشرون درهما فتجمع تركة العبد وهي

و خمسون درهما غير شيئين وسدس شيء وهو مثلا الوصيتين جميعا التين هما شيئان وثلثا شيء فاجبر ذلك فيكون ثماني ماية وخمسين درهما يعدل سبعة اشياء ونصفا فقابل به فيكون الشيء الواحد يعدل ماية وثلثة عشر درهما وثلث درهم وذلك وصية العبد الذي قيمته ثلثماية درهم و وصية العبد الاخر مثل ذلك ومثل ثلثيه وذلك ماية وثمانية وثمانون درهما وثمانية اتساع درهم وسعايته ثلثماية وأحد عشر درهما وتسع درهم *

فان اعتق عبدين له في مرضه قيمة كل واحد منهما ثلثماية درهم ثم مات احدهما وترك خمسماية درهم وترك بنتا وترك السيد ابنا فقياسه ان تجعل وصية كل واحد منهما شيئا و سعايته ثلثماية غير شيء و تجعل تركة الميت منهما خمسماية درهم وسعايته ثلثماية غير شيء فيبقي ما ترك مايتان وشيء فيرجع الي مولاه بالميراث ماية درهم ونصف شيء فيصير في ايدي ورثة مولاه اربعماية درهم غير نصف شيء و ياخذون من العبد الاخر سعايته ثلثماية درهم غير شيء فيصير في ايديهم سبعماية درهم ونصف شيء فذلك مثلا صيتهما التي هي الشيئان وذلك اربعة اشياء فاجبر ذلك بشيء

بقي من الماية ويسعي الاخر في مايعين و ثلثة وثلثين درهما و ثلث *

فان اعتق عبدين له في مرضه قيمة احدهما ثلثماية درهم و قيمة الاخر خمسماية درهم فمات الذي قيمته ثلثماية درهم وترك بنتا وترك السيد ابنا وترك العبد اربعماية درهم في كم يسعي كل واحد منهما فقياسه ان تجعل وصية العبد الذي قيمته ثلثماية درهم شيئا و سعايته ثلثماية غير شيء و تجعل وصية العبد الذي قيمته خمسماية درهم شيئا و ثلثي شيء و سعايته خمسماية درهم غير شيء و ثلثي شيء لان قيمته مثل قيمة الاول ومثل ثلثيها فاذا كان لذلك شيء كان لهذا مثله و مثل ثلثيه فمات الذي قيمته ثلثماية درهم و ترك اربعماية درهم تودي من ذلك السعاية ثلثماية غير شيء فيبقي في ايدي ورثته ماية درهم وشيء النصف من ذلك لابنته وهو خمسون درهما و نصف شيء وما بقي لورثة السيد وهو خمسون درهما و نصف شيء مضاف الي ثلثماية غير شيء فيكون ثلثماية و خمسين غير نصف شيء و ياخذون من الاخر سعايته وهو خمسماية درهم غير شيء و ثلثي شيء فيصير في ايديهم ثماني ماية

عشرون درهما و تسعا شيء فيصير في ايدي ورثة المولي ثلثماية وعشرون غير سبعة اتساع شيء يقضي من ذلك دين المولي عشرون درهما فيبقي ثلثماية غير سبعة اتساع شيء وذلك مثلا ما كان للعبد من الوصية التي هي شيء وذلك شيئان فتجبر الثلثماية بسبعة اتساع شيء تزيد ذلك علي الشيئين فيبقي ثلثماية يعدل شيئين وسبعة اتساع شيء الشيء من ذلك تسعة اجزاء من خمسة و عشرين فيكون ذلك ماية وثمانية و ذلك ما كان للعبد *

فان اعتق عبدين له في مرضه ولا مال له غيرهما وقيمة كل واحد منهما ثلثماية درهم فتعجل المولي من احدهما ثلثي قيمته فاستهلكها ثم مات السيد فماله ثلث قيمة الذي تعجل منه فمال السيد جميع قيمة الذي لم يتعجل منه وثلث قيمة الذي تعجل منه وهو ماية درهم و ذلك اربع ماية درهم وثلث ذلك بينهما نصفان وهو ماية درهم وثلثة وثلثون درهما وثلث درهم لكل واحد منهما ستة وستون درهما و ثلثا درهم فيسعي الذي تعجل منه ثلثي قيمته في ثلثة وثلثين درهما وثلث لان له من الماية ستة وستين درهما وثلثي درهم وصية ويسعي فيما

دراهم من ذلك وصية المرأة شيء فيبقي ماية درهم و عشرة دراهم غير شيء و يصير في ايدي ورثة المرأة عشرون درهما وشيء واوصت من ذلك بثلثه وهو ستة دراهم وثلثان و ثلث شيء ويرجع الي ورثة الزوج من ذلك بالميراث نصف ما بقي وهو ستة دراهم وثلثان وثلث شيء فيصير في ايدي ورثة الزوج ماية وستة عشر درهما وثلثان غير ثلث شيء واوصي من ذلك بثلثه وهو شيء فيبقي ماية درهم وستة عشر درهما وثلثان غير شيء وثلثي شيء يعدل مثلي الوصيتين وذلك اربعة اشياء فاجبر ذلك فيكون ماية و ستة عشر درهما و ثلثي درهم يعدل خمسة اشياء وثلثي شيء فالشيء الواحد يعدل عشرين درهما وعشرة اجزاء من سبعة عشر جزءا من درهم وهي الوصية فاعلم ذلك *

---

### باب العتق في المرض *

---

اذا اعتق الرجل عبدين له في مرضه و ترك السيد ابنا وابنة ثم مات احد العبدين و ترك مالا اكثر من قيمته و ترك ابنة فاجعل ثلثي قيمته وما سعي فيه العبد الاخر وميراث السيد منه بين الابن والبنت للذكر مثل حظ

فان كان تزوجها علي ماية درهم و مهر مثلها عشرة دراهم واوصى لرجل بثلث ماله فقياس ذلك ان تعطي المرأة مهرها وهو عشرة دراهم فيبقي تسعون درهما ثم تعطي من ذلك وصيتك شيئا ثم تعطي الموصي له بالثلث ايضا شيئا لان الثلث بينهما نصفان لا تاخذ المرأة شيئا الا اخذ صاحب الثلث مثله فتعطي صاحب الثلث ايضا شيئا ثم يرجع الي ورثة الزوج ميراثه من المرأة خمسة دراهم و نصف شيء فيبقي في ايدي ورثة الزوج خمسة وتسعون الا شيء و نصفا وذلك يعدل اربعة اشياء فاجبر ذلك بشيء و نصف شيء فيبقي خمسة و تسعون يعدل خمسة اشياء و نصف فاجعلها انصافا فيكون احد عشر نصفا والدراهم انصافا فتكون ماية و تسعين نصفا يعدل احد عشر شيئا فالشيء الواحد يعدل سبعة عشر درهما و ثلثة اجزاء من احد عشر من درهم فهي الوصية *

فان تزوجها علي ماية درهم و مهر مثلها عشرة دراهم ثم ماتت قبل الزوج و تركت عشرة دراهم واوصت بثلث مالها ثم مات الزوج و ترك ماية وعشرين درهما واوصي لرجل بثلث ماله فقياسه ان تعطي المرأة مهرها عشرة دراهم فيبقي في ايدي ورثة الزوج ماية درهم وعشرة

## باب التكملة

امرأة ماتت و تركت ثماني بنات و امها و زوجها واوصت لرجل بتكملة خمس المال بنصيب بنت ولاخر بتكملة ربع المال بنصيب الام فقياس ذلك ان تقيم سهام الفريضة فيكون ثلثة عشر سهما فتاخذ مالا فتلقي منه خمسه الا سهما نصيب بنت وهي الوصية الاولي ثم تلقي منه ايضا ربعه الا سهمين نصيب الام وهي الوصية الثانية فيبقي احد عشر جزءا من عشرين جزءا من مال وثلثة اسهم يعدل ثلثة عشر سهما فالت من الثلثة عشر السهم ثلثة اسهم بثلثة اسهم فيبقي معك احد عشر جزءا من عشرين من مال يعدل عشرة اسهم فكمل مالك وهو ان تزيد علي العشرة الاسهم تسعة اجزاء من احد عشر جزءا منها فيكون معك مال يعدل ثمانية عشر سهما وجزؤين من احد عشر جزءا من سهم فاجعل السهم احد عشر فيكون المال مايتين والسهم احد عشر والوصية الاولي تسعة وعشرون والثانية ثمانية وعشرون *

فان كانت الفريضة علي حالها واوصت لرجل بتكملة الثلث بنصيب الزوج ولاخر بتكملة الربع بنصيب الام

سهما من مايتين واربعين سهما من مال واربعة اخماس نصيب ودرهم واربعة اخماس درهم فخذ الثلث وهو ثمانون فالق منه اثني عشر واربعة اخماس نصيب ودرهما واربعة اخماس درهم ثم الق ربع ما بقي معك ودرهما فيبقي معك من الثلث احد وخمسون الا ثلثة اخماس نصيب والا درهمين وسبعة اجزاء من عشرين جزءا من درهم ثم الق من ذلك ثمن المال وهو ثلثون فيبقي احد وعشرون الا ثلثة الاخماس نصيب والا درهمين وسبعة اجزاء من عشرين جزءا من درهم وثلثا المال يعدل ثمانية انصبا فاجبر ذلك بما نقص وزده علي الثمانية الانصبا فيكون معك ماية واحد وثمانون سهما من مال يعدل ثمانية انصبا وثلثة اخماس نصيب ودرهمين وسبعة اجزاء من عشرين جزءا من درهم وكمل مالك وذلك ان تزيد علي ما معك تسعة وخمسين من ماية واحد وثمانين فيكون النصيب ثلثماية واثنين وستين والدرهم ثلثماية واثنين وستين والمال خمسة الاف ومايتين وستة وخمسين والوصايا من الربع الف ومايتان واربعة ومن الثلث اربعماية وتسعة وتسعون والثمن ستماية وسبعة وخمسون *

واربعة اخماس نصيب فيبقي خمسة غير اربعة اخماس نصيب فتلق ربع ذلك ايضا للوصية و درهما فيبقي معك سهمان و ثلثة ارباع سهم الا ثلثة اخماس نصيب ثم الق ثمن المال وهو ثلثة فيبقي عليك بعد الثلث ربع سهم و ثلثة اخماس نصيب فارجع الي الثلثين وهما ستة عشر فالق من ذلك ربع واحد و ثلثة اخماس نصيب فيبقي من المال خمسة عشر سهما و ثلثة ارباع سهم غير ثلثه اخماس نصيب [يعدل ثمانية انصبا] فاجبر ذلك بثلثة اخماس نصيب وزدها علي الانصبا وهي ثمانية فيكون خمسة عشر سهما و ثلثة ارباع سهم يعدل ثمانية انصبا و ثلثة اخماس نصيب فاقسم ذلك عليه فما بلغ فهو القسم وهو النصيب والمال اربعة و عشرون و يكون لكل بنت سهم و ماية و ثلثة واربعون جزءا من ماية واثنين وسبعين جزءا من سهم * فان اردت ان تخرج السهام صحيحة فخذ ربع مال فالق منه نصيبا فيبقي ربع مال الا نصيبا ثم الق منه درهما ثم الق خمس ما بقي من الربع وهو خمس ربع مال الا خمس نصيب والا خمس درهم والق درهما ثانيا فيبقي اربعة اخماس الربع الا اربعة اخماس نصيب والا درهما و اربعة اخماس درهم غالوصية من الربع اثني عشر

ودرهما وثلثي درهم فكمل مالك وهو ان تزيد علي الاربعة الانصبا والخمسة الاسداس و الدرهم وثلثي الدرهم جزءا من سبعة عشر جزءا من نصيب ودرهما وثلثي عشر جزءا من سبعة عشر جزءا من درهم فاجعل النصيب سبعة عشر سهما و الدرهم سبعة عشر فيكون المال ماية وسبعة عشر وان اردت ان تخرج الدرهم صحيحا فاعمل به كما وصفت لك ان شاء الله تعالي *

فان ترك ثلثة بنين وابنتين واوصي لرجل بمثل نصيب بنت وبدرهم ولاخر بخمس ما بقي من الربع وبدرهم ولاخر بربع ما بقي من الثلث بعد ذلك كله وبدرهم ولاخر بثمن جميع المال فاجاز ذلك الورثة فقياسه علي ان تخرج الدراهم صحاحا وهو في هذا الوجه احسن هو ان تاخذ ربع مال و تسميه [فاجعله] ستة والمال اربعة و عشرين فالق من الربع نصيبا فيبقي ستة غير نصيب ثم الق درهما فيبقي خمسة غير نصيب فالق خمس ما يبقي فيبقي اربعة غير اربعة اخماس نصيب ثم الق درهما اخر فيبقي معك ثلثة غير اربعة اخماس نصيب فقد علمت ان الوصية من الربع ثلثة و اربعة اخماس نصيب ثم ارجع الي الثلث فالق منه ثلثة

فما بلغ فهو القسم وهو النصيب وهو ثلثة وجزء من احد عشر من درهم والثلث سبعة ونصف  *

فان ترك اربعة بنين واوصى لرجل بمثل نصيب احد بنيه الا ربع ما يبقى من الثلث بعد النصيب وبدرهم ولاخر بثلث ما يبقى من الثلث وبدرهم فان الوصية من الثلث فخذ ثلث مال فالق منه نصيبا فيبقى ثلث الا نصيبا ثم زد علي ما معك ربعه فيكون ثلثا وربع ثلث الا نصيبا وربع نصيب والق درهما فيبقى ثلث وربع ثلث الا درهما والا نصيبا وربع نصيب ثم الق ثلث ما يبقى معك من الوصية الثانية فيبقى معك من الثلث خمسة اسهم من ستة اسهم من ثلث مال الا ثلثى درهم والا خمسة اسداس نصيب ثم الق درهما اخر فيبقى معك خمسة اسهم من ثمانية عشر سهمها من مال الا درهما وثلثي درهم والا خمسة اسداس نصيب فزد علي ذلك ثلثي المال فيكون معك سبعة عشر سهما من ثمانية عشر سهما من مال الا درهما وثلثي درهم والا خمسة اسداس نصيب يعدل اربعة انصبا فاجبر ذلك بها نقص وزد مثله علي الانصبا فيكون سبعة عشر سهما من ثمانية عشر من مال يعدل اربعة انصبا وخمسة اسداس نصيب

خمسة انصبا فاجبر ذلك بنصف نصيب وبدرهم وثلثة
ارباع درهم وزدها علي الانصبا فيكون معك خمسة اسداس
مال تعدل خمسة انصبا ونصف نصيب ودرهما وثلثة
ارباع درهم فكمل مالك وهو ان تزيد علي الانصبا والدرهم
والثلثة الارباع مثل خمسها فيكون معك مال يعدل ستة
انصبا و ثلثة اخماس نصيب و درهمين و عشر درهم
فاجعل النصيب عشرة و الدرهم عشرة فيكون المال سبعة
وثمانين سهما * وان اردت ان تخرج الدرهم درهما
صحيحا فخذ الثلث فاطرح منه نصيبا فيكون ثلثا الا نصيبا
واجعل الثلث سبعة ونصفا ثم الق ثلث ما معك وهو
ثلث الثلث فيبقي معك ثلثا الثلث الا ثلثي نصيب
وهو خمسة دراهم الا ثلثي نصيب فالق واحدا بالدرهم
فيبقي معك اربعة دراهم الا ثلثي نصيب ثم الق ربع
ما معك وهو سهم الا السدس نصيب والق سهما بالدرهم
فيبقي معك سهمان الا النصف نصيب فزد ذلك علي ثلثي
المال وهو خمسة عشر فيكون سبعة عشر الا نصف نصيب
يعدل خمسة انصبا فاجبر ذلك بنصف نصيب وزده علي
الخمسة فيكون سبعة عشرسهما يعدل خمسة انصبا ونصفا
فاقسم سبعة [عشر] علي خمسة انصبا ونصف نصيب

ودرهما وجزءا من احد عشر من درهم * فان اردت ان تخرج الدرهم صحيحا فلا تكمل مالك فلكن اطرح من الاحد عشر واحدا بالدرهم واقسم العشرة الباقية علي الانصبا اربعة انصبا وهي اربعة و ثلثة ارباع نصيب فيكون القسم اثنين وجزءا من تسعة عشر اجزاء من درهم فاجعل المال اثني عشر والنصيب سهمين وجزوين من تسعة عشر جزءا وان اردت ان تخرج النصيب صحيحا فتتمّم مالك واجبره فيكون الدرهم احد عشر من المال *

فان ترك خمسة بنين واوصي لرجل بمثل نصيب احدهم وبثلث ما يبقي من الثلث و بدرهم و بربع ما يبقي بعد ذلك من الثلث و بدرهم فخذ ثلثا فالق منه نصيبا فيبقي ثلث الا نصيبا ثم الق ما يبقي معك وهو ثلث الثلث الا ثلث نصيب ثم الق مما يبقي درهما فيبقي معك ثلثا الثلث الا ثلثي نصيب والا درهما ثم الق مما معك ربعه وهو سهم من ستة اسهم من الثلث الا سدس نصيب و الا ربع درهم ثم الق درهما اخر يبقي معك نصف الثلث الا نصف نصيب والا درهما و ثلثة ارباع درهم فزد علي ذلك ثلثي المال فيكون خمسة اسداس مال الا نصف نصيب والا درهما وثلثة ارباع درهم يعدل

تسعة واربعون والوصية من الربع عشرة والمستثني من النصيب الثاني ستة فافهم ذلك *

## باب الوصية بالدرهم *

رجل مات وترك اربعة بنين واوصى لرجل بمثل نصيب احدهم و بربع ما بقي من الثلث و بدرهم فقياس ذلك ان تاخذ ثلث مال فتلقي منه نصيبا فيبقي ثلث الا نصيبا ثم تلقي ربع ما يبقي معك وهو ربع ثلث الا ربع نصيب و تلقي ايضا درهما فيبقي معك ثلثة ارباع ثلث مال وهو ربع المال الا ثلثة ارباع نصيب والا درهما فتزيد ذلك علي ثلثي المال فيكون معك احد عشر جزءا من اثني عشر من مال الا ثلثة ارباع نصيب والا درهما يعدل اربعة انصبا فاجبر ذلك بثلثة ارباع نصيب و بدرهم فيكون احد عشر جزءا من اثني عشر من مال يعدل اربعة انصبا و ثلثة ارباع نصيب ودرهما فكمل مالك وهو ان تزيد علي الانصبا والدرهم جزءا من احد عشر جزءا منها فيكون معك مال يعدل خمسة انصبا و جزءين من احد عشر جزءا من نصيب

والنصيب الاخر فان قياسه ان تلقي من ربع مال نصيبا فيبقي ربع غير نصيب ثم تلقي خمس ما يبقي من الربع وهو نصف عشر المال الا خمس نصيب ثم ترجع الي الثلث فتلقي منه نصف عشر المال و اربعة اخماس نصيب ونصيبا اخر فيبقي ثلث الا نصف عشر المال والا نصيبا واربعة اخماس نصيب فزد علي ذلك ربع ما يبقي وهو الذي استثناه فاجعل الثلث ثمانين فاذا رفعت نصف عشر المال بقي منه ثمانية وستون الا نصيبا واربعة اخماس نصيب فزد علي ذلك ربعه وهو سبعة عشر سهما الا ربع ما تنقص من الانصبا فيكون ذلك خمسة وثمانين الا نصيبين وربع نصيب فزد ذلك علي ثلثي المال وهو ماية وستون فيكون معك مال وسدس ثمن مال الا نصيبين وربعا يعدل ستة انصبا فاجبر ذلك بما نقص منه وزده علي الانصبا فيكون مالا وسدس ثمن مال يعدل ثمانية انصبا وربع نصيب فاردد ذلك الي مال واحد وهو ان تنقص من الانصبا جزءا من تسعة واربعين جزءا من جميعها فيكون مال يعدل ثمانية انصبا واربعة اجزاء من تسعة واربعين جزءا من نصيب فاجعل النصيب تسعة واربعين فيكون المال ثلثماية وستة وتسعين والنصيب

تاخذ ايضا ربع مال فتلقي منه نصيبا فيبقى معك ربع مال غير نصيب ثم تلقي ثلث ما يبقى من الربع فيبقي ثلثا ربع الا ثلثي نصيب فتزيد ذلك علي ما يبقى من الثلث فيكون ذلك ستة و عشرين جزءا من ستين جزءا من مال غير نصيب وثمانية و عشرين جزءا من ستين جزءا من نصيب ثم زد علي ذلك ما بقي من المال بعد اخذك منه الثلث والربع وهو ربع و سدس فيكون ذلك سبعة عشر جزءا من عشرين جزءا من مال يعدل سبعة انصبا و سبعة اجزاء من خمسة عشر جزءا من نصيب فتمم مالك وهو ان تزيد علي ما معك من الانصبا ثلثة اجزاء من سبعة عشر جزءا فيكون معك مال يعدل ثمانية انصبا و ماية و عشرين جزءا من ماية و ثلثة و خمسين جزءا من نصيب فاجعل النصيب ماية و ثلثة و خمسين فيكون المال الفا و ثلثماية واربعة واربعين والوصية من الثلث بعد النصيب تسعة و خمسون والوصية من الربع بعد النصيب احد وستون

فان ترك ستة بنين واوصي لرجل بمثل نصيب ابن وبخمس ما يبقى من الربع و لرجل اخر بمثل نصيب ابن اخر ربع الا ربع ما يبقى من الثلث بعد الوصيتين الوليين

وخمس نصيب ثم تلقي من ذلك نصيب بنت اخري فيبقي ثلث و خمس ثلث الا نصيبين وخمس نصيب ثم تزيد علي ذلك ما استثني فيكون ثلثا وثلثة اخماس ثلث الا نصيبين واربعة عشر جزءا من خمسة عشر جزءا من نصيب ثم تلقي من ذلك نصف سدس جميع المال فيبقي سبعة و عشرون جزءا من ستين من مال الا ما ينقص من الانصبا فزد علي ذلك ثلثي المال و اجبره بما نقص من الانصبا وزدها علي الانصبا فيكون معك مال و سبعة اجزاء من ستين جزءا من مال يعدل ثمانية انصبا و اربعة عشر جزءا من خمسة عشر جزءا من نصيب فاردد ذلك الي مال واحد وهو ان تنقص مما معك سبعة اجزاء من سبعة و ستين منه فيكون النصيب مايتين و واحدا و يصير المال كله الفا و ستماية و ثمانية *

فان كانت الفريضة علي حالها واوصي بمثل نصيب بنت وبخمس ما يبقي من الثلث بعد النصيب و بمثل نصيب بنت اخري و بثلث ما يبقي من الربع بعد نصيب واحد فقياس ذلك ان الوصيتين من الربع ومن الثلث فتاخذ ثلث مال فتلقي منه نصيبا فيبقي ثلث مال الا نصيبا ثم تلقي خمس ما يبقي وهو خمس ثلث الا الخمس نصيب فيبقي . اربعة اخماس ثلث الا اربعة اخماس نصيب ثم

تسعة اجزاء من تسعة و خمسين جزءا فيبقي مال يعدل ثمانية انصبا وثلثة وعشرين جزءا من تسعة وخمسين جزءا من نصيب فالنصيب تسعة وخمسون جزءا وتكون سهام الفريضة اربعماية وخمسة وتسعين سهما والخمسان من ذلك ماية وثمانية وتسعون سهما فارفع من ذلك النصيبين ماية وثمانية عشر سهما يبقي ثمانون سهما ترفع منه المستثني وهو ربع الثمانين وخمسها ستة وثلثون سهما فيبقي للموصي له اثنان وثمانون سهما ترفعها من سهام الفريضة وهي اربعماية وخمسة وتسعون سهما فيبقي اربعماية وثلثة عشر سهما بين سبعة انصبا لكل بنت تسعة وخمسون وللابن مثل ذلك *

---

فان ترك ابنين وابنتين واوصي لرجل بمثل نصيب بنت الا الخمس ما يبقي من الثلث بعد النصيب ولاخر بمثل نصيب بنت اخري الا ثلث ما يبقي من الثلث بعد ذلك كله واوصي لرجل اخر بنصف سدس جميع المال فان هذه الوصايا كلها من الثلث فتاخذ ثلث مال فتلقي منه نصيب بنت فيبقي ثلث مال الا نصيبا ثم تزيد علي ذلك ما استثني وهو خمس الثلث الا خمس نصيب فيكون ذلك ثلثا وخمس ثلث الا نصيبا

وخمسة وخمسين والخمسان من ذلك ثلثماية واثنان ثم ارفع النصيب من ذلك وهو اثنان وثمانون فيبقي مايتان وعشرون ثم ارفع من ذلك الربع والخمس تسعة وتسعين سهما فتبقي ماية واحد وعشرون فزد عليها ثلثة اخماس المال وهو اربعماية وثلثة وخمسون فيكون خمسماية واربعة وسبعين بين سبعة اسهم لكل سهم اثنان وثمنون وهو نصيب البنت وللابن ضعف ذلك *

فان كانت الفريضة علي حالها واوصي لرجل بمثل نصيب الابن الا ربع وخمس ما يبقي من الخمسين بعد النصيب فالوصية من الخمسين ترفع من ذلك نصيبين لان للابن سهمين فيبقي خمسا مال الا نصيبين وزد ما استثنا عليه وهو ربع الخمسين وخمسها الا تسعة اعشار نصيب فيكون خمس مال وتسعة اعشار الخمس الا نصيبين وتسعة اعشار نصيب فزد علي ذلك ثلثة اخماس المال فيكون مالا وتسعة اعشار خمس مال الا نصيبين وتسعة اعشار نصيب يعدل سبعة انصبا فاجبر ذلك بنصيبين وتسعة اعشار نصيب وزدها علي الانصبا فيكون معك مال وتسعة اعشار خمس مال يعدل تسعة انصبا وتسعة اعشار نصيب فاردد ذلك الي مال واحد وهو ان تنقص مما معك

بين سبعة اسهم لكل سهم ماية وثمانية وثمانون سهما وهو نصيب البنت وللابن ضعف ذلك *

فان كانت الفريضة علي حالها واوصي من خمسي ماله بمثل نصيب البنت ولآخر بربع وخمس ما يبقي من الخمسين بعد النصيب فقياس ذلك ان الوصية من الخمسين فتاخذ خمسي مال فتلقي منه النصيب فيبقي خمسا مال الا نصيبا ثم تلقي منه ربع وخمس ما يبقي وهو تسعة اجزاء من عشرين جزءا من الخمسين الا مثل ذلك من النصيب فيبقي خمس وعشر الخمس الا احد عشر جزءا من عشرين جزءا من نصيب فزد عليه ثلثة اخماس المال فيكون ذلك اربعة اخماس وعشر خمس مال الا احد عشر جزءا من عشرين جزءا من نصيب يعدل سبعة انصبا فاجبر ذلك باحد عشر جزءا من عشرين جزءا من نصيب وزدها علي السبعة فيكون ذلك يعدل سبعة انصبا واحد عشر جزءا من عشرين جزءا من نصيب فتمم مالك وهو ان تزيد علي كل ما معك تسعة اجزاء من احد واربعين جزءا فيكون معك مال يعدل تسعة انصبا وسبعة عشر جزءا من اثنين وثمانين جزءا من نصيب فاجعل النصيب اثنين وثمانين جزءا فيكون السهام سبعماية

نصيب ابنة فاطرح منه الوصية الاخري وهي خمسة وسدسه فيبقي سبع واربعة اجزاء من خمسة عشر جزءا من سبع الا تسعة عشر جزءا من ثلثين جزءا من نصيب فزد علي ذلك خمسة اسباع المال الباقية فيكون ستة اسباع مال واربعة اجزاء من خمسة عشر من سبع المال الا تسعة عشر جزءا من ثلثين جزءا من نصيب يعدل سبعة انصبا فاجبرها بتسعة عشر جزءا وزدها علي السبعة الانصبا فيكون ستة اسباع مال واربعة اجزاء من خمسة عشر جزءا من سبع مال يعدل سبعة انصبا وتسعة عشر جزءا من ثلثين جزءا من نصيب فكمل مالك وهو ان تزيد علي كل ما معك احد عشر جزءا من اربعة وتسعين جزءا فيكون معك مال يعدل ثمانية انصبا وتسعة وتسعين جزءا من مايه وثمانية وثمانين جزءا من نصيب فاجعل المال كله الفا وستمايه وثلثة والنصيب ماية وثمانية وثمانين ثم خذ سبعي المال وهو اربعماية وثمانية وخمسون فاطرح منه النصيب وهو ماية وثمانية وثمانون ويبقي مايتان وسبعون فاطرح خمس ذلك وسدسه تسعة وتسعين سهما فيبقي ماية واحد وسبعون سهما فزد عليه خمسة اسباع المال وهو الف وماية وخمسة واربعون فيكون الفا وثلثماية وستة عشر سهما

ثم اردد اليه نما استثني وهو خمس الثلث الا خمس نصيب فيكون ثلثا و خمس ثلث وذلك خمسان الا نصيبا و خمس نصيب ثم زد ذلك علي ثلثي المال فيكون مالا و خمس ثلث مال الا نصيبا و خمس نصيب يعدل اربعة انصبا فاجبر المال بنصيب و خمس نصيب وزده علي الاربعة الانصبا فيكون مالا و خمس ثلث مال يعدل خمسة انصبا و خمس نصيب فاردد ذلك الي مال واحد وهو ان تنقص مما معك نصف ثمنه وهو جزء من ستة عشر فيصير معك مال يعدل اربعة انصبا و سبعة اثمان نصيب فاجعل المال تسعة و ثلثين والمال ثلثة عشر والنصيب ثمانية فيبقي من الثلث خمسة خمسها واحد فزد عليه الواحد الذي استثناه من الوصية فتبقي الوصية سبعة ويبقي من الثلث ستة فزد عليها ثلثي المال وهو ستة و عشرون سهما فيكون اثنين و ثلثين علي اربعة بنين لكل ابن ثمانية *

فان ترك ثلثة بنين وبنتا واوصي لرجل من سبعي ماله بمثل نصيب ابنته وللاخر بخمس وسدس ما يبقي من السبعين فالوصية في هذا الوجه من سبعي المال فخذ سبعي المال فاطرح منه نصيب ابنة فيبقي سبعا مال الا

المال في هذا النوع وقياسه ان تاخذ ثلث مال فتلقي منه النصيب فيبقي ثلث مال الا نصيبا ثم تنقص منه ربع ما يبقي من الثلث وهو ربع ثلث الا ربع نصيب فيبقي ربع مال الا ثلثة ارباع نصيب فزد عليه ثلثي المال فيكون احد عشر جزءا من اثني عشر جزءا من مال الا ثلثة ارباع نصيب يعدل اربعة انصبا فاجبر ذلك بثلثة ارباع نصيب وزدها علي الاربعة الانصبا فيكون معك احد عشر جزءا من اثني عشر جزءا من مال يعدل اربعة انصبا وثلثة ارباع نصيب فكمل مالك وهو ان تزيد علي الاربعة الانصبا والثلثة الارباع جزءا من احد عشر جزءا فيكون ذلك خمسة انصبا وجزؤين من احد عشر من نصيب يعدل مالا فاجعل النصيب احد عشر والمال سبعة و خمسين والثلث تسعة عشر ترفع ذلك النصيب احد عشر فيبقي منه ثمانية للموصي له بالربع اثنان ويبقي ستة مردودة علي الثلثين وهما ثمانية وثلثون فيكون اربعة واربعين بين اربعة بنين لكل ابن احد عشر سهما *

فان تركت اربعة بنين واوصي لرجل بمثل نصيب ابن الا خمس ما يبقي من الثلث بعد النصيب فالوصية من الثلث فخذ ثلثا واطرح منه نصيبا فيبقي ثلث الا نصيبا

جزءًا من ماية و تسعة اجزاء من سهم فتجعل السهم ماية و تسعة اجزاء و تضرب الثلثة عشر في ماية و تسعة اجزاء و تزيد علي ذلك ثمانين جزءًا فيكون الفا واربعماية وسبعة و تسعين و نصيب الزوج ثلثماية و سبعة و عشرون *

فان ترك اختين وامرأة واوصي لرجل بمثل نصيب اخت الا ثمن ما يبقي من المال بعد الوصية فقياس ذلك ان تقيم الفريضة من اثني عشر سهما لكل اخت ثلث ما يبقي من المال بعد الوصية فهذا مال الا وصية فانت تعلم ان ثمن ما يبقي مع الوصية يعدل نصيب اخت فثمن ما يبقي هو ثمن مال الا ثمن وصية فثمن مال الا ثمن وصية مع وصية يعدل نصيب اخت و ذلك ثمن مال و سبعة اثمان وصية فالمال كله يعدل ثلثة اثمان مال وثلث وصايا وخمسة اثمان وصية فاطرح من المال ثلثة اثمانه فيبقي خمسة اثمان المال تعدل ثلثة وصايا وخمسة اثمان وصية فالمال كله يعدل خمس وصايا واربعة اخماس وصية فالمال تسعة وعشرون والوصية خمسة والنصيب ثمانية *

وفي وجه اخر من الوصايا رجل مات و ترك اربعة بنين واوصي لرجل بمثل نصيب احد بنيه ولخر بربع ما يبقي من الثلث فاعلم ان الوصية انما هي من ثلث

احد وثلثين منها وهي ماية واربعة واربعون جزءا فيكون ذلك ستماية واربعين فالثمنها وعشرها ماية واربعة واربعين ومثل نصيب الزوج وهو ثلثة وتسعون فيبقي اربعماية وثلثة للزوج من ذلك ثلثة وتسعون وللام اثنان وستون ولكل بنت ماية واربعة وعشرون *

فان كانت الفريضة علي حالها واوصت لرجل بمثل نصيب الزوج الا تسع وعشر ما يبقي من المال بعد النصيب فقياس ذلك ان تقيم سهام الفريضة فتخذها من ثلثة عشر سهما والوصية من جميع المال ثلثة اسهم فيبقي مال الا ثلثة اسهم ثم استثني تسع وعشر ما يبقي من المال فهو تسع مال وعشره الا تسع ثلثة اسهم وعشرها وذلك تسعة عشر جزءا من ثلثين جزءا من سهم فيكون ذلك مالا وتسعا وعشرا الا ثلثة اسهم وتسعة عشر جزءا من ثلثين من سهم يعدل ثلثة عشر سهما فاجبر مالك بثلثة اسهم وتسعة عشر جزءا من سهم فزده علي الثلثة عشر مثلها فيكون مالا وتسعا وعشرا يعدل ستة عشر سهما وتسعة عشر جزءا من ثلثين جزءا من سهم فرد ذلك الي مال واحد وهو ان تنقص من ذلك تسعة عشر جزءا من ماية وتسعة اجزاء فيبقي مال يعدل ثلثة عشر سهما وثماني

ثلثة عشر سهما للام من ذلك سهمان وانت تعلم ان الوصية سهمان وتسع جميع المال فيبقي منه ثمانية اتساع المال الا سهمين بين الورثة فتتمم مالك وتمامه ان تجعل الثمانية الاتساع الا سهمين ثلثة عشر سهما فتزيد علي ذلك سهمين فيكون خمسة عشر سهما يعدل ثمانية اتساع مال ثم تزيد علي ذلك ثمنه وعلي خمسة عشر ثمنها وهو سهم وسبعة اثمان سهم لصاحب التسع من ذلك التسع وهو سهم وسبعة اثمان سهم وللاخر الموصي له بمثل نصيب الام سهمان فيبقي ثلثة عشر سهما بين الورثة علي سهامهم وتصح من ماية وخمسة وثلثين سهما *

فان اوصت بمثل نصيب الزوج وبثمن المال وعشرة فاقم سهام الفريضة فتكون ثلثة عشر سهما ثم زد عليها مثل نصيب الزوج وهو ثلثة فيكون ستة عشر وذلك ما بقي من المال بعد الثمن والعشر وهو تسعة اجزاء من اربعين سهما والذي يبقي من المال بعد الثمن والعشر احد وثلثون جزءًا من اربعين جزءًا من مال وهو يعدل ستة عشر سهما فكمل مالك وهو ان تزيد عليه تسعة اجزاء من احد وثلثين جزءًا فاضرب ستة عشر في احد وثلثين منها فيكون ذلك اربعماية وستة وتسعين فزد عليها تسعة اجزاء من

نصيب ابن وثلثي ما بقي من الثلث فنحذ ثلثا فاطرح منه اربعة اسباع نصيب ابن فيبقي ثلث مال الا اربعة اسباع نصيب ابن ثم الق ثلث ما بقي من الثلث وهو تسع مال الا سبع نصيب وثلث سبع نصيب فيبقي تسع مال الا سبعي نصيب وثلثي سبع نصيب فزد ذلك علي ثلثي المال فيكون ثمانية اتساع مال الا سبعي نصيب وثلثى سبع نصيب و ذلك ثمانية اجزاء من واحد وعشرين جزءا من نصيب تعدل ثلثة انصبا فاجبر ذلك فيكون ثمانية اتساع مال تعدل ثلثة انصبا وثمانية اجزاء من احد وعشرين جزءا من نصيب فتمم مالك وهو ان تزيد علي الثمانية الاتساع مثل ثمنها و علي الانصبا مثل ثمنها فيكون معك مال يعدل ثلثة انصبا وخمسة واربعين جزءا من ستة وخمسين جزءا من نصيب والنصيب ستة و خمسون والمال مايتان وثلثة عشر سهما والوصية الاولي اثنان وثلثون سهما والثانية ثلثة عشر و بقي ماية وثمانية وستون لكل ابن ستة وخمسون سهما   *

و في وجه اخر من الوصايا   *   امرأة ماتت و تركت ابنتيها وامها وزوجها واوصت لرجل بمثل نصيب الام ولاخر بتسع جميع المال فقياس ذلك تقيم سهام الفريضة فتكون

البنون ثلثة كم كانت تكون سهامهم فتاخذ ذلك سبعة
فيخذ فريضة يكون لخمسها سبع ولسبعها خمس وذلك
خمسة وثلثون فزد عليه سبعيها وهو عشرة فيكون ذلك
خمسة واربعين للموصي له من ذلك عشرة ولكل ابن اربعة
عشر وللبنت سبعة *

فان ترك أُمّا وثلثة بنين وبنتا واوصي لرجل بمثل
نصيب احد بنيه الا مثل نصيب بنت اخري لو كانت
فاقم سهام الفريضة واجعلها شيئا ينقسم بين هولاء الورثة
وبينهم لو كانت معهم ابنة اخري فتاخذها ثلثماية وستة
وثلثين فنصيب ابنة لو كانت خمسة وثلثون ونصيب
ابن ثمانون سهما وبينهما خمسة واربعون وهي الوصية فزدها
علي ثلثماية وستة وثلثين فيكون ذلك ثلثماية وأحدا
وثمانين فذلك سهام المال *

فان ترك ثلثة بنين واوصي لرجل بمثل نصيب احد
البنين الا مثل نصيب ابنة لو كانت وبثلثي ما بقي من
الثلث فقياس ذلك ان تقيم سهام الفريضة علي شيء
ينقسم بين هولاء الورثة وبينهم لو كانت معهم ابنة اخري
فيكون ذلك واحدا وعشرين فلو كانت معهم بنت اخري
لكان لها ثلثة ونصيب ابن سبعة فقد اوصي له باربعة اسباع

فصل ما بين خمسي نصيبه وبين ما نصيبه من الثلث وهو ثمانية وثلثون من ماية وخمسة وتسعين من نصيب الابن بعد اخراج الثلث لهما لان الذي له من حاصة الثلث ثمانية اجزاء من ثلثة عشرة من الثلث وهو اربعون والذي اجاز له من خمسي نصيبه ثمانية وثلثون فذلك ثمانية وسبعون فيوخذ منه خمسة وستون ثلث ماله لهما والذي اجاز له حاصة ثمانية وثلثون فان اردت تصحيح سهام الفريضة صححتها فكانت من مايتي الف وتسعة عشر الفا وثلثماية وعشرين   *

وفي وجه اخر من الوصايا رجل مات وترك اربعة بنين وامرأة واوصي لرجل بمثل نصيب احد البنين الا مثل نصيب المرأة فاقم سهام الفريضة وهي اثنان وثلثون سهما للمرأة الثمن اربعة ولكل ابن سبعة فانت تعلم ان الذي اوصي له به ثلثة اسباع نصيب ابن فزد علي الفريضة ثلثة اسباع نصيب ابن وهو ثلثة وهي الوصية فيكون ذلك خمسة وثلثين للموصي له ثلثة اسهم من خمسة وثلثين سهما فيبقي اثنان وثلثون بين الورثة علي سهامهم   *

فان تركت ابنين وبنتا واوصي لرجل بمثل نصيب ابن ثالث لو كان فالوجه في ذلك ان تنظر الي ابن لو كان

لهما فاضرب سهام الفريضة في ثلثة عشر يصح من ثلثة الاف وماية وعشرين
*

فان اجاز الابن الخمسين لصاحب الخمسين ولم يجز للاخر شيئا واجازت الام الربع لصاحب الربع ولم يجز للاخر شيئا ولم يجز الزوج لهما الا الثلث فاعلم ان الثلث للرجلين جائز علي جميع الورثة يضرب فيه صاحب الخمسين بثمانية اجزاء من ثلثة عشر جزءا وصاحب الربع بخمسة اجزاء من ثلثة عشر فاقم الفريضة علي ما ذكرت لك فيكون اثني عشر للزوج الربع وللام السدس وللابن ما بقي وقياسه انك تعلم ان الزوج يخرج من يده ثلث حصته علي كل حال فينبغي ان يكون في يده ثلثة اسهم وان الام يخرج من يدها الثلث لكل واحد بقدر حصته وهي قد اجازت لصاحب الربع من حاصه حصتها فصل ما بين الربع وحصته من نصيبها وهي تسعة عشر جزءا من ماية وستة وخمسين من جميع نصيبها فينبغي ان يكون نصيبها ماية وستة وخمسين فحصته من الثلث من نصيبها عشرون سهما والذي اجازت له ربع حصتها وهو تسعة وثلثون وتوخذ ثلث ما في يدها لهما وتسعة عشر سهما للذي اجازت له حاصة ثم الابن قد اجاز لصاحب الخمسين

الفريضة فتاخذها من اثني عشر سهما للابن من ذلك سبعة اسهم وللزوج ثلثة اسهم وللام سهمان * وانت تعلم ان الزوج يجوز عليه الثلث فينبغي ان يكون في يده مثلا ما يخرج من حصته للوصايا و في يده ثلثة للوصايا سهم وله سهمان * واما الابن الذي اجاز الوصيتان جميعا فينبغي ان يوخذ منه خمسا جميع ماله وربعه فيبقي في يده سبعة اسهم من عشرين سهما والذي له كله عشرون سهما * واما الام فينبغي ان يبقي في يدها مثل ما يخرج من يدها وهو واحد وجميع ما كان لها اثنان * فخذ مالا يكون لاربعه ثلث ولسدسه نصف ويكون ما يبقي يتقسم بين عشرين فذلك مايتان واربعون * للام من ذلك السدس وهو اربعون الوصية من ذلك عشرون ولها عشرون * وللزوج من ذلك الربع ستون الوصية من ذلك عشرون وله اربعون * ويبقي ماية واربعون للابن الوصية من ذلك خمسان وربعة وهو واحد و تسعون ويبقي تسعة واربعون فجميع الوصية ماية واحد و ثلثون بين الرجلين الموصي لهما لصاحب الخمس من ذلك ثمانية اجزاء من ثلثة عشر جزءا ولصاحب الربع خمسة اجزاء من ثلثة عشر جزءا فان اردت تصحيح سهام الرجلين الموصي

فتاخذها من عشرين فتاخذ مالا فالتي ثمنه وسبعه فيبقي مال الا ثمنا وسبعا فتتم مالك وهو ان تزيد عليه خمسة عشر جزءا من احد واربعين جزءا فاضرب سهام الفريضة وهي عشرون في احد واربعين فيكون ثماني ماية وعشرين فتزيد علي ذلك خمسة عشر جزءا من احد واربعين وهو ثلثماية جزء فيصير ذلك كله الفا وماية وعشرين سهما للموصي له من ذلك بالثمن والسبع سبع ذلك وثمنه وهو ثلثماية السبع ماية وستون والثمن ماية واربعون فيبقي ثماني ماية وعشرون سهما بين الورثة علي سهامهم *

---

باب اخر من الوصايا *

وهو اذا لم يجز بعض الورثة واجاز بعضهم والوصية اكثر من الثلث * اعلم ان الحكم في ذلك ان من اجاز من الورثة اكثر من الثلث من الوصية فذلك داخل عليه في حصته ومن لم يجز فالثلث جائز عليه علي كل حال * مثال ذلك امرأة ماتت وتركت زوجها وابنها وامها واوصت لرجل بخمسي مالها ولاخر بربع مالها فاجاز الابن الوصيتين جميعا واجازت الام النصف لهما ولم يجز الزوج شيئا من ذلك الا الثلث فقياس ذلك ان تقيم سهام

جزءا من شيء يعدل ثلثة دراهم فتحتاج الي ان تكمل الشيء فتزيد عليه اربعة اجزاء من احد عشر من شيء وتزيد مثل ذلك علي ثلثة دراهم وهو درهم و جزؤ من احد عشر جزءا فيكون اربعة دراهم وجزءا من احد عشر جزءا من درهم يعدل شيئا وهو الذي استخرج من الدين *

---

باب اخر من الوصايا *

---

رجل مات وترك امه وامرأته واخاه واختيه لابيه وامه واوصي لرجل بتسع ماله فان قياس ذلك ان تقيم فريضتهم فتاخذها من ثمانية و اربعين سهما فانت تعلم ان كل مال نزعت تسعه بقيت ثمانية اتساعه وان الذي نزعت مثل ثمن ما ابقيت فتزيد علي الثمانية الاتساع ثمنها وعلي الثمانية والاربعين مثل ثمنها ليتم مالك وهو ستة فيكون ذلك اربعة وخمسين للموصي له بالتسع من ذلك ستة وهو تسع جميع المال وما بقي فهو ثمانية واربعون بين الورثة علي سهامهم *

فان قال امرأة هلكت وتركت زوجها وابنها وثلث بنات واوصت لرجل بثمن مالها وسبعه فاقم سهام الفريضة

بخمس ماله وهو درهمان وخمس شيء فيبقي ثمانية دراهم واربعة اخماس شيء ثم تعزل الدرهم الذي اوصي به فيبقي سبعة دراهم واربعة الخماس شيء فتقسمه بين الابنين فيكون لكل واحد ثلثة دراهم ونصف درهم وخمسا شيء [وهو يعدل الشيء فقابل به فتلقي خمسي شيء] من شيء فيبقي ثلثة اخماس شيء تعدل ثلثة دراهم ونصفا فكمل الشيء وهو ان تزيد عليه مثل ثلثيه وتزيد علي الثلثة والنصف مثل ثلثيه وهو درهمان وثلث فيكون خمسة وخمسة اسداس وهو الشيء الذي استخرج من الدين *

فان ترك ثلثة بنين واوصي بخمس ماله الا درهما وترك عشرة دراهم عينا وعشرة دراهم دينا علي احد البنين فان قياسه ان تجعل المستخرج من الدين شيئا فتزيده علي العشرة فيكون عشرة وشيئا فتعزل خمسها للوصية وهو درهمان وخمس شيء فيبقي ثمانية دراهم واربعة اخماس شيء ثم تستثني درهما لانه قال الا درهما فيكون تسعة دراهم واربعة اخماس شيء فتقسم ذلك بين البنين فيكون لكل ابن ثلثة دراهم وخمس شيء وثلث خمس شيء فيكون ذلك يعدل شيئا فتلقي خمس شيء وثلث خمس شيء من شيء فيبقي احد عشر جزءا من خمسة عشر

كتاب الوصايا *

باب من ذلك في العين والدين *

رجل مات وترك ابنين واوصى بثلث ماله لرجل اخر وترك عشرة دراهم عينا وعشرة دراهم دينا علي احد الابنين فقياسه ان تجعل المستخرج من الدين شيئا فتزيده علي العين وهو عشرة دراهم فيكون عشرة وشيئًا ثم تعزل ثلثها لانه اوصى بثلث ماله وهو ثلثة دراهم وثلث وثلث شيء فيبقي ستة دراهم وثلثان وثلثا شيء فتقسمه بين الابنين فنصيب كل ابن ثلثة دراهم وثلث درهم وثلث شيء فهو يعدل الشيء المستخرج فقابل به فتلقي ثلثا من شيء بثلث شيء فيبقي ثلثا شيء يعدل ثلثة دراهم وثلثا فتحتاج ان تكمل الشيء [فتزيد عليه مثل نصفه وتزيد علي الثلثة والثلث مثل نصفها فيكون خمسة دراهم وهي الشيء] الذي استخرج من الدين *

فان ترك ابنين وترك عشرة دراهم عينا وعشرة دراهم دينا علي احد الابنين واوصى لرجل بخمس ماله ودرهم فقياسه ان تجعل ما يستخرج من الدين شيئا فتزيده علي العين فيكون شيئا وعشرة دراهم فتعزل خمسها لانه اوصى

K

العمود وتكسيرها ثمانية واربعون ذراعا وهو ضربك العمود في نصف القاعدة وهو ستة فجعلنا احد جوانب المربعة شيئا فضربناه في مثله فصار مالا فحفظناه ثم علمنا انه قد بقي لنا مثلثان عن جنبتي المربعة ومثلثة فوقها فاما المثلثان اللتان علي جنبتي المربعة فهما متساويتان وعموداهما واحد وهما علي زاوية قائمة فتكسيرها ان تضرب شيئا في ستة الا نصف شيء فيكون ستة اشياء الا نصف مال وهو تكسير المثلثين جميعا اللتان هما علي جنبتي المربعة فاما تكسير المثلثة العليا فهو ان تضرب ثمانية غير شيء وهو العمود في نصف شي فيكون اربعة اشياء الا نصف مال فجميع ذلك هو تكسير المربعة وتكسير الثلث المثلثات وهو عشرة اشياء تعدل ثمانية واربعين هو تكسير المثلثة العظمي فالشيء الواحد من ذلك اربعة اذرع واربعة اخماس ذراع وهو كل جانب من المربعة   *   وهذه صورتها   *

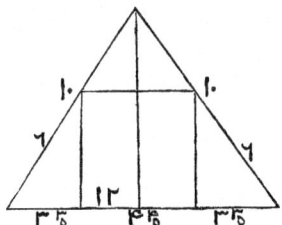

وهو عشرون ذراعا فبلغ ذلك ماية وستة اذرع وثلثي ذراع فاردنا ان نلقي منه ما زدنا عليه حتي يخرط وهو واحد وثلث الذي هو ثلث تكسير اثنين في اثنين في عشرة وهو ثلثة عشر وثلث وذلك تكسير ما زدنا عليه حتي انخرط فاذا رفعنا ذلك من ماية وستة اذرع وثلثي ذراع بقي ثلثة و تسعون ذراعا و ثلث وذلك تكسير العمود المخروط وهذه صورته *

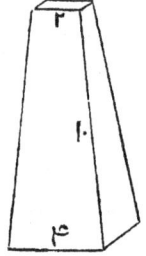

وإن كان المخروط مدورا فالق من ضرب قطره في نفسه سبعه ونصف سبعه فما بقي فهو تكسيره *

فان قيل ارض مثلثة من جانبيه عشرة اذرع عشرة اذرع والقاعدة اثنا عشر ذراعا في جوفها ارض مربعة كم كل جانب من المربعة فقياس ذلك ان تعرف عمود المثلثة وهو ان تضرب نصف القاعدة وهو ستة في مثله فيكون ستة وثلثين فانقصها من احد الجانبين الاقصرين مضروبا في مثله وهو ماية يبقي اربعة وستون فخذ جذرها ثمانية وهو

الكتاب فمنها مدورة قطرها سبعة اذرع ويحيط بها اثنان وعشرون ذراعا فان تكسيرها ان تضرب نصف القطر وهو ثلثة ونصف في نصف الدور الذي يحيط بها وهو احد عشر فيكون ثمانية وثلثين ونصفا وهو تكسيرها فان احببت فاضرب القطر وهو سبعة في مثله فيكون تسعة واربعين فانقص منها سبعها ونصف سبعها وهو عشرة ونصف فيبقي ثمانية وثلثون ونصف وهو التكسير وهذه صورتها *

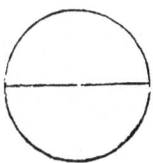

فان قال عمود مخروط اسفله اربعة اذرع في اربعة اذرع وارتفاعه عشرة اذرع وراسه ذراعان في ذراعين وقد بيّنا ان كل مخروط محدد الراس فان ثلث تكسير اسفله مضروبا في عموده هو تكسيره فلما صار هذا غير محدد اردنا ان نعلم كم يرتفع حتى يكمل رأسه فيكون لا رأس له فعلمنا ان هذه العشرة من الطول كله كعد الاثنين من الاربعة فالاثنان نصف الاربعة فاذا كان ذلك كذلك فالعشرة نصف الطول والطول كله عشرون ذراعا فلما عرفنا الطول اخذنا ثلث تكسير الاسفل وهو خمسة وثلث فضربناه في الطول

وهو اثني عشر والعمود ابدا يقع علي القاعدة علي زاويتين
قائمتين ولذلك سمي عمودا لانه مستو فاضرب العمود في
نصف القاعدة وهو سبعة فيكون اربعة وثمانين وذلك
تكسيرها وذلك صورتها  *

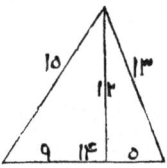

والجنس الثالث منفرجة وهي التي لها زاوية منفرجة
وهي مثلث من كل جانب عدد مختلف وهي من
جانب ستة ومن جانب خمسة ومن جانب تسعة فمعرفة
تكسير هذه من قبل عمودها ومسقط حجرها ولا يقع مسقط
حجر هذه المثلثة في خوفها الا علي الضلع الاطول فاجعله قاعدة
ولو جعلت احد الضلعين الاقصرين قاعدة لوقع مسقط حجرها
خارجها وعلم مسقط حجرها وعمودها علي مثال ما علمتك
في الحادة وعلي ذلك القياس وهذه صورتها  *

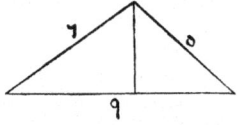

واما المدورات التي فرغنا من صفتها وتكسيرها في صدر

منها علي شيء مما يلي اي الضلعين شئت فجعلنا
الشيء مما يلي الثلثة عشر فضربناه في مثله فصار مالا
ونقصناه من ثلثة عشر في مثلها وهو ماية و تسعة وستون
فصار ذلك ماية و تسعة وستين الا مالا فعلمنا ان جذرها
هو العمود وقد بقي لنا من القاعدة اربعة عشر الا شيئا
فضربناه في مثله فصار ماية وستة و تسعين ومالا الا ثمانية
و عشرين شيئا فنقصناه من الخمسة عشر في مثلها فبقي
تسعة وعشرون درهما وثمانية وعشرون شيئا وجذرها
هو العمود فلما صار جذرها هذا هو العمود وجذر ماية
وتسعة وستين الا مالا هو العمود ايضا علمنا انهما متساويان
فقابل بهما وهو ان تلقي مالا بمال لان المالين ناقصان
فيبقي تسعة وعشرون وثمانية وعشرون شيئا يعدل ماية
و تسعة و ستين فالق تسعة و عشرين من ماية وتسعة
وستين فيبقي ماية واربعون يعدل ثمانية و عشرين شيئا
فالشيء الواحد خمسة وهو مسقط الحجر مما يلي الثلثة
عشر و تمام القاعدة مما يلي الضلع الاخر فهو تسعة فاذا
اردت ان تعرف العمود فاضرب هذه الخمسة في مثلها
وانقصها من الضلع الذي يليها مضروبا في مثله وهو ثلثة
عشر فيبقي ماية و اربعة و اربعون فجذر ذلك هو العمود

مبلغ الخمسة في مثلها وهو خمسة وعشرون فيبقى خمسة وسبعون فخذ جذر ذلك فهو العمود وقد صار ضلعا على مثلثتين قائمتين فان اردت التكسير فاضرب جذر الخمسة والسبعين في نصف القاعدة وهو خمسة وذلك ان تضرب الخمسة في مثلها حتى تكون جذر خمسة وسبعين في جذر خمسة وعشرين فاضرب خمسة وسبعين في خمسة وعشرين فيكون الفا وثماني ماية وخمسة وسبعين فخذ جذر ذلك وهو تكسيرها وهو ثلثة واربعون وشيء قليل وهذه صورتها *

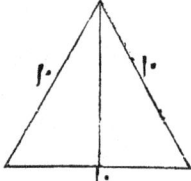

وقد تكون من هذه العادة الزوايا مختلفة الاضلاع فاعلم ان تكسيرها يعلم من قبل مسقط حجرها وعمودها وهي ان تكون مثلثة من جانب خمسة عشر ذراعا ومن جانب اربعة عشر ذراعا ومن جانب ثلثة عشر ذراعا فاذا اردت علم مسقط حجرها فاجعل القاعدة اي الجوانب شئت فجعلناها اربعة عشر وهو مسقط الحجر فمسقط حجرها يقع

منها ستة اذرع و ضلع منها ثمانية اذرع و القطر عشر فحساب ذلك ان تضرب ستة في اربعة فيكون اربعة و عشرين ذراعا وهو تكسيرها * وان احببت ان تحسبها بالعمود فان عمودها لا يقع الا علي الضلع الاطول لان الضلعين القصيرين عمودان فان اردت ذلك فاضرب عمودها في نصف القاعدة فما كان فهو تكسيرها وهذه صورتها *

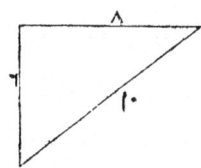

واما الجنس الثاني فالمثلثة المتساوية الاضلاع حادة الزوايا من كل جانب عشرة اذرع فان تكسيرها تعرف من قبل عمودها ومسقط حجرها واعلم ان كل ضلعين متساويين من مثلثة تخرج منهما عمود علي قاعدة فان مسقط حجر العمود يقع علي زاوية قائمة ويقع علي نصف القاعدة سوا اذا استوا الضلعان فان اختلفا خالف مسقط الحجر عن نصف القاعدة ولكن قد علمنا ان مسقط حجر هذه المثلثة علي اي اضلاعها جعلته لا يقع الا علي نصفه فذلك خمسة اذرع فمعرفة العمود ان تضرب الخمسة في مثلها وتضرب احد الضلعين في مثله وهو عشر فيكون ماية فتنقص منها

فيخرج الى حساب المثلثات فاعلم ذلك وهذه صورة المشبهة بالمعينة *

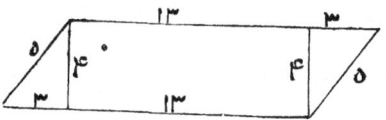

واما المثلثات فهي ثلثة اجناس القائمة والحادة والمنفرجة * واما القائمة فهي مثلثة اذا ضربت ضلعيها الاقصرين كل واحد منهما في نفسه ثم جمعتهما [كان مجموع ذلك مثل الذي يكون من ضرب الضلع الاطول في نفسه * واما الحادة فهي مثلثة اذا ضربت ضلعيها الاقصرين كل واحد منهما في نفسه ثم جمعتهما] كانا اكثر من الضلع الاطول مضروبا في نفسه * واما المنفرجة فهي كل مثلثة اذا ضربت ضلعيها الاقصرين كل واحد منهما في نفسه وجمعتهما كانا اقل من الضلع الاطول مضروبا في نفسه *

فاما القائمة الزوايا فهي التي لها عمودان وقطر وهي نصف مربعة فمعرفة تكسيرها ان تضرب احد الضلعين المحيطين بالزاوية القائمة في نصف الاخر فما بلغ ذلك فهو تكسيرها * ومثل ذلك مثلثة قائمة الزاوية ضلع

٥٦

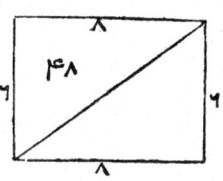

و اما المعينة المستوية الاضلاع التي كل جانب منها خمسة اذرع فاحد قطريها ثمانية والاخر ستة اذرع فاعلم ان تكسيرها ان تعرف القطرين او احدهما فان عرفت القطرين جميعا فان الذي يكون من ضرب احدهما في نصف الاخر هو تكسيرها وذلك ان تضرب ثمانية في ثلثة او اربعة في ستة فيكون اربعة وعشرين ذراعا وهو تكسيرها فان عرفت قطرا واحدا فقد علمت انهما مثلثان كل واحد منهما ضلعاها خمسة اذرع خمسة اذرع والضلع الثالث هو قطرهما فاحسبهما على حساب المثلثات وهذه صورتها *

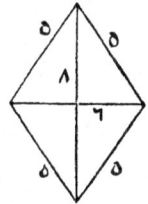

واما المشبهة بالمعينة فعلي مثل المعينة *
واما سائر المربعات فانما تعرف تكسيرها من قبل القطر

اعلم ان المربعات خمسة اجناس فمنها مستوية الاضلاع قائمة الزوايا والثانية قائمة الزوايا مختلفة الاضلاع طولها اكثر من عرضها والثالثة تسمى المعينة وهي التي استوت اضلاعها واختلفت زواياه والرابعة المشبهة بالمعينة وهي التي طولها وعرضها مختلفان وزواياها مختلفة غير ان الطولين مستويان والعرضين مستويان ايضا والخامسة المختلفة الاضلاع والزوايا *

فما كان من المربعات مستوية الاضلاع قائمة الزوايا او مختلفة الاضلاع قائمة الزوايا فان تكسيرها ان تضرب الطول في العرض فما بلغ فهو التكسير * ومثال ذلك ارض مربعة من كل جانب خمسة اذرع تكسيرها خمسة وعشرون ذراعا وهذه صورتها *

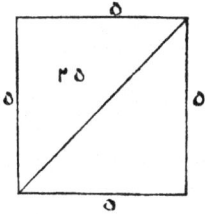

والثانية ارض مربعة طولها ثمانية اذرع ثمانية اذرع والعرضان ستة ستة فتكسيرها ان تضرب ستة في ثمانية فيكون ثمانية واربعين ذراعا وذلك تكسيرها وهذه صورتها *

الي نقطة ط خطا يقطع سطح اك بنصفين فحدث من السطح مثلثين وهما مثلثا اطه و هكط فقد تبين لنا ان اط نصف اب و اه مثله وهو نصف اج وتوترهما خط طه علي زاوية قائمة وكذلك نخرج خطوطا من ط الي ر ومن ر الي ح ومن ح الي ه فحدث من جميع المربعة ثماني مثلثات متساويات وقد تبين لنا ان اربع منها نصف السطح الاعظم الذي هو اد وقد تبين لنا ان خط اط في نفسه تكسير مثلثين و اه تكسير مثلثين بمثلهما فيكون جميع ذلك تكسير اربع مثلثات وضلع هط في نفسه ايضا تكسير اربع مثلثات اخر وقد تبين لنا ان الذي يكون من ضرب اط في نفسه و اه في نفسه مجموعين مثل الذي يكون من ضرب طه في نفسه وذلك ما اردنا ان نبين وهذه صورته *

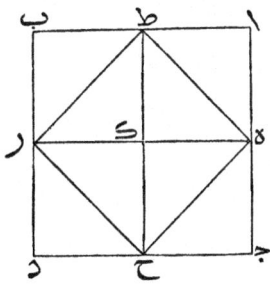

حفظت ان كانت القوس اقل من نصف مدورة او زده عليه ان كانت القوس اكثر من نصف مدورة فما بلغ بعد الزيادة او النقصان فهو تكسير القوس      *

وكل مجسم مربع فان ضربك الطول في العرض ثم في العمق هو التكسير  *  فان كان علي غير تربيع وكان مدورا او مثلثا او غير ذلك الا ان عمقه علي الاستواء والموازاة فان مساحة ذلك ان تمسح سطحه فتعرف تكسيره فما كان ضربته في العمق وهو التكسير     *

واما المخروط من المثلث والمربع والمدور فان الذي يكون من ضرب ثلث مساحة اسفله في عموده هو تكسيره      *

واعلم ان كل مثلث قائم الزاوية فان الذي يكون من ضرب الضلعين الاقصرين كل واحد منهما في نفسه مجموعيين مثل الذي يكون من ضرب الضلع الاطول في نفسه      *

وبرهان ذلك انا نجعل سطحا مربعا متساوي الاضلاع والزوايا اب جد ثم نقتطع ضلع اج بنصفين علي نقطة ه ثم نخرجه الي ز ثم نقتطع ضلع اب بنصفين علي نقطة ط ونخرجه الي نقطة ح فصار سطح اب جد اربعة سطوح متساوية الاضلاع والزوايا والمساحة وهي سطح اك وسطح جك وسطح بك وسطح دك ثم نخرج من نقطة ه

من المثلثات والمربعات والمخمسات وما فوق ذلك فان ضربك نصف ما يحيط بها في نصف قطر اوسع دائرة يقع فيها تكسيرها * و كل مدورة فان قطرها مضروبا في نفسه منقوصا منه سبعه و نصف سبعه هو تكسيرها وهو موافق للباب الاول *

و كل قطعة من مدورة مشبهة بقوس فلا بد ان يكون مثل نصف مدورة او اقل من نصف مدورة او اكثر من نصف مدورة و الدليل علي ذلك ان سهم القوس اذا كان مثل نصف الوتر فهي نصف مدورة سوا واذا كان اقل من نصف الوتر فهي اقل من نصف مدورة واذا كان السهم اكثر من نصف الوتر فهي اكثر من نصف مدورة * واذا اردت ان تعرف من اي دائرة هي فاضرب نصف الوتر في مثله و اقسمه علي السهم وزد ما خرج علي السهم فما بلغ فهو قطر المدورة التي تلك القوس منها * فان اردت ان تعرف تكسير القوس فاضرب نصف قطر المدورة في نصف القوس واحفظ ما خرج ثم انقص سهم القوس من نصف قطر المدورة ان كانت القوس اقل من نصف مدورة وان كانت اكثر من نصف مدورة فانقص نصف قطر المدورة من سهم القوس ثم اضرب ما بقي في نصف وتر القوس وانقصه مما

مثل ربع السطح الذي هو من كل جانب ذراع وكذلك ثلث في ثلث وربع في ربع وخمس في خمس وثلثان في نصف او اقل من ذلك او اكثر فعلي حسابه * وكل سطح مربع متساوي الاضلاع فان احد اضلاعه في واحد جذره وفي اثنين جذراه صغر ذلك السطح او كثر *

وكل مثلث متساوي الاضلاع فان ضربك العمود و نصف القاعدة التي يقع عليها العمود هو تكسير ذلك المثلث *

وكل معينة متساوية الاضلاع فان ضربك احد القطرين في نصف الاخر هو تكسيرها *

وكل مدورة فان ضربك القطر في ثلثة وسبع هو الدور الذي يحيط بها وهو اصطلاح بين الناس من غير اضطرار * ولاهل الهندسة فيه قولين اخران احدهما ان تضرب القطر في مثله ثم في عشرة ثم تاخذ جذر ما اجتمع فما كان فهو الدور * فالقول الثاني لاهل النجوم منهم وهو ان تضرب القطر في اثنين وستين الفا وثماني ماية واثنين وثلثين ثم تقسم ذلك علي عشرين الفا فما خرج فهو الدور وكل ذلك قريب بعضه من بعض * والدور اذا قسمته علي ثلثة وسبع يخرج القطر * وكل مدورة فان نصف القطر في نصف الدور هو التكسير لان كل ذات اضلاع وزوايا متساوية

عمل بستة ايام كم نصيبه فقد علمت ان الستة الايام هي خمس الشهر وان الذي نصيبه من الدراهم بقدر ما عمل من الشهر وقياس ذلك ان قوله شهر هو ثلثون يوما وهو المسعر وقوله عشرة دراهم هو السعر و قوله ستة ايام هو المثمن وقوله كم نصيبه هو الثمن فاضرب السعر الذي هو عشرة في المثمى الذي هو مبائنه وهو ستة فيكون ستين فاقسمه علي الثلثين التي هي العدد الظاهر وهو المسعر فيكون ذلك درهمين وهو الثمن وهذا ما يتعامل الناس بينهم من الصرف والكيل والوزن *

## باب المساحة *

اعلم ان معني واحد في واحد انما هي مساحة ومعناه ذراع في ذراع * وكل سطح متساوي الاضلاع والزوايا يكون من كل جانب واحد فان السطح كله واحد * فان كان من كل جانب اثنان ،هو متساوي الاضلاع والزوايا فالسطح كله اربعة امثال السطح الذي هو ذراع في ذراع * وكذلك ثلثة في ثلثة وما زاد علي ذلك او نقص وكذلك نصف في نصف بربع وغير ذلك من الكسور فعلي هذا * وكل سطح مربع يكون من كل جانب نصف ذراع فهو

لك باربعة فقوله عشرة هو العدد المسعر وقوله بستة هو السعر وقوله كم لك هو العدد المجهول المثمن وقوله باربعة هو العدد الذي هو الثمن فالعدد المسعر الذي هو العشرة هو مباين للعدد الذي هو الثمن وهو الاربعة فاضرب العشرة في الاربعة وهما المتباينان الظاهران فيكون اربعين فاقسمها علي العدد الاخر الظاهر الذي هو السعر وهو ستة فيكون ستة وثلثين وهو العدد المجهول الذي هو في قول القائل كم وهو المثمن ومباينه الستة الذي هو السعر *

والوجه الثاني قول القائل ثمانية عشرة كم ثمن اربعة وربما قال اربعة منها كم ثمنها فالعشرة هي العدد المسعر وهو مباين للعدد الذي هو الثمن المجهول الذي في قوله كم والثمانية هي العدد الذي هو السعر وهو مباين للعدد الظاهر الذي هو المثمن وهو اربعة فاضرب العددين الظاهرين المتباينين احدهما في الاخر وهو اربعة في ثمانية فيكون اثنين وثلثين واقسمه علي العدد الاخر الظاهر الذي هو المسعر وهو عشرة فيكون ثلثة و خمسا وهو العدد الذي هو الثمن وهو مباين للعشرة التي عليها قسمت وهكذا جميع معاملات الناس وقياسها ان شاء الله تعالي *

فان سأل سائل فقال اجير اجرته في الشهر عشرة دراهم

فان قال قال مال تعزل ثلثة اجذاره ثم تضرب ما بقي في مثله فيعود المال فقد علمت ان الذي بقي هو جذر ايضا والمال اربعة اجذار وهو ستة عشر *

باب المعاملات *

اعلم ان معاملات الناس كلها فمن البيع والشري والصرف والاجارة و غير ذلك علي وجهين باربعة اعداد يلفظ بها السائل وهي المسعر والسعر والثمن والمثمن فالعدد الذي هو المسعر مبائن للعدد الذي هو المثمن والعدد الذي هو السعر مبائن للعدد الذي هو الثمن وهذه الاربعة الاعداد ثلثة منها ابدا ظاهرة معلومة و واحد منها مجهول وهو الذي في قول القايل كم و عنه يسأل السائل * والقياس في ذلك ان تنظر الي الثلثة الاعداد الظاهرة فلا بد ان يكون منها اثنان كل واحد منهما مبائن لصاحبه فتضرب العددين الظاهرين المتبائنين كل واحد منهما في صاحبه فما بلغ فاقسمه علي العدد الاخر الظاهر الذي مبائنه مجهول فما خرج لك فهو العدد المجهول الذي يسأل عنه السائل فهو مبائن للعدد الذي قسمت عليه *

ومثال ذلك في وجه منه اذا قيل لك عشرة بستة كم

المال الاول كله من قبل ان تلقي ثلثيه في ثلثة اجذاره كان مالا و نصفا لان ثلثيه في ثلثة اجذاره مال فهو كله في ثلثة اجذاره مال و نصف وهو كله في جذر واحد نصف مال فجذر المال نصف والمال ربع فثلثا المال سدس و ثلثة اجذار المال درهم و نصف فمتى ما ضربت سدسا في درهم و نصف خرج ربعا وهو المال         *

فان قال مال تعزل اربعة اجذاره ثم تاخذ ثلث ما بقي فيكون مثل الاربعة الاجذار والمال مايتان و ستة و خمسون فقياسه انك تعلم ان ثلث ما بقي مثل الاربعة الاجذار وان بقي مثل اثني عشر جذره فزد عليه الاربعة الاجذار فيكون ستة عشر جذرا وهو جذر المال       *

فان قال مال عزلت جذره وزدت علي جذره جذر ما بقي فكان درهمين فهذا جذر مال فجذر مال الا جذرا يعدل درهمين فالق منه جذر مال والق من الدرهمين جذر مال فيكون درهمين الا جذرا في مثله اربعة دراهم ومالا الا اربعة اجذار يعدل مالا الا جذرا فقابل به فيكون مالا واربعة دراهم يعدل مالا وثلثة اجذار فتلقى مالا بمال فيبقي ثلثة اجذار تعدل اربعة دراهم فالجذر يعدل درهما وثلثا وهو جذر المال والمال درهم وسبعة اتساع درهم     *

مال وسدس جذر مقسوم علي درهم يعدل درهما فكمل المال الذي معك وهو ان تضربه في ستة فيكون معك مال وجذر فاضرب الدرهم في ستة فيكون ستة دراهم فيكون مالا وجذرا يعدل ستة دراهم فنصف الجذر واضربه في مثله فيكون ربعا فزده علي الستة و خذ جذر ما اجتمع فانقص منه نصف الجذر الذي كنت ضربته في مثله وهو نصف ما بقي فهو عدد الرجال الوليين وهم في هذه المسئلة رجلان *

فان قال مال ضربته في ثلثيه فكان خمسة فقياسه انك اذا ضربته في مثله كان سبعة و نصفا فتقول هو جذر سبعة و نصف في ثلثي جذر سبعة و نصف فاضرب ثلثين في ثلثين فيكون اربعة اتساع واربعة اتساع في سبعة و نصف يكون ثلثة و ثلثا فجذر ثلثة و ثلث هو ثلثا جذر سبعة و نصف فاضرب ثلثة و ثلثا في سبعة و نصف فيكون خمسة و عشرين فجذرها خمسة * فان قال مال تضربه في ثلثة اجذاره فيكون خمسة امثال المال الاول فكانه قال مال ضربته في جذره فكان مثل المال الاول و ثلثيه فجذر المال درهم و ثلثان والمال درهمان وسبعة اتساع *

فان قال مال تلقي ثلثيه ثم تضرب الباقي في ثلثة اجذار المال الاول فيعود المال الاول وقياسه انك اذا ضربت

تضرب شيئا في ثلثي شيء فيكون ثلثي مال يعدل خمسة فاكمله بمثل نصفه وزد علي الخمسة مثل نصفها فيصير معك مال يعدل سبعة و نصفا فخذ جذرها وهو الشيء الذي تريد ان تضربه في ثلثيه فيكون خمسة   *

فان قال مالان بينهما درهمان قسمت القليل علي الكثير فاصاب القسم نصف درهم فقياسه ان تضرب شيئا ودرهمين في القسم وهو نصف فيكون نصف شيء ودرهما يعدل شيئا فالقي نصف شيء بنصف شيء يبقي درهم يعدل نصف شيء فاضعفه فيكون معك شيء يعدل درهمين وهو احد المالين والمال الاخر اربعة   *

فان قال قسمت درهما علي رجال فاصابهم شيء ثم زدت فيهم رجالا ثم قسمت عليهم درهما فاصابهم اقل من القسم الاول بسدس درهم فقياسه ان تضرب عدد الرجال الاولين وهم شيء في النقصان الذي بينهم ثم تضرب ما اجتمع في عدد الرجال الاولين و الاخرين ثم تقسم ما اجتمع علي ما بين الرجال الاولين والاخرين فانه يخرج مالك الذي قسمته فاضرب عدد الرجال الاولين وهو شيء في السدس الذي بينهم فيكون سدس جذر ثم اضرب ذلك في عدد الرجال الاولين والاخرين وهو شيء و واحد يكون سدس

و تضرب الاربعة الدراهم في خمسة و تسعة عشر جزءًا من خمسة و عشرين فيكون ثلثة و عشرين درهمًا وجزءًا من خمسة و عشرين و تضرب اربعة اجذار و ثلثا في خمسة و تسعة عشر جزءًا من خمسة وعشرين فيكون اربعة وعشرين جذرًا و اربعة و عشرين جزءًا من خمسة و عشرين من جذر فنصف الاجذار فتكون اثني عشر جذرًا واثني عشر جزءًا من خمسة و عشرين من جذر واضربها في مثلها فيكون ماية و خمسة و خمسين درهمًا و اربعماية و تسعة وستين جزءًا من ستماية وخمسة وعشرين فالق منها الدراهم الثلثة و العشرين والجزء من الخمسة والعشرين الذي كان مع المال فتبقي ماية واثنان وثلثون واربعماية و اربعون جزءًا من ستماية و خمسة و عشرين فتاخذ جذر ذلك وهو احد عشر درهمًا وثلثة عشر جزءًا من خمسة وعشرين فتزيده علي نصف الاجذار التي هي اثني عشر درهمًا واثني عشر جزءًا من خمسة وعشرين فيكون ذلك اربعة وعشرين وهو المال المطلوب الذي تعزل ثلثه وربعه واربعة دراهم ثم تضرب ما بقي في مثله فيعود المال وزيادة اثني عشر درهمًا *

فان قال مال ضربته في ثلثيه فبلغ خمسة فقياسه ان

خمسة اجزاء من اثني عشر من شيء الا اربعة دراهم فتضربها في مثلها فتكون الاجزاء الخمسة خمسة و عشرين جزءا فتضرب الاثني عشر في مثلها فيكون ماية واربعة واربعين فذلك خمسة و عشرون من ماية واربعة واربعين من مال ثم تضرب الاربعة الدراهم في الخمسة الاجزاء من اثني عشر من شيء مرتين فيكون اربعين جزءا كل اثني عشر منها شيء والاربعة الدراهم والاربعة الدراهم ستة عشر درهما زايدة فتصير الاربعون الجزء ثلثة اجذار و ثلث جذر ناقص فيحصل معك خمسة وعشرون جزءا من ماية واربعة واربعين جزءا من مال و ستة عشر درهما الا ثلثة اجذار وثلث جذر يعدل المال الاول وهو شيء واثني عشر درهما فاجبره وزد الثلثة الاجذار والثلث علي الشي و الاثني عشر درهما فيصير اربعة اجذار و ثلث جذر و اثني عشر درهما فقابل به والق اثني عشر من ستة عشر يبقي اربعة دراهم و خمسة و عشرون جزءا من ماية واربعين من مال يعدل اربعة اجذار وثلثا فيحتاج ان تكمل مالك و اكمالك اياه ان تضرب جميع ما معك في خمسة و تسعة عشر جزءا من اجزاء خمسة و عشرين فتضرب خمسة و عشرين في خمسة و تسعة عشر جزءا من خمسة و عشرين فيكون مالا

جزءا من جذر يعدل جذرا وثلثة عشر درهما فالق درهمين من ثلثة عشر بدرهمين فيبقي احد عشر درهما فالق احد عشر جزءا من جذر فيبقي نصف سدس جذر واحد عشر درهما يعدل نصف سدس مال فاكمله وذلك ان تضربه في اثني عشر و تضرب كل ما معك في اثني عشر فيكون مالا يعدل ماية و اثنين و ثلثين درهما و جذرا فقابل به يصب ان شاء الله تعالي كما وصفت لك *

فان قال درهم و نصف مقسوم علي رجل وبعض رجل فاصاب الرجل مثل البعض فقياسه ان تقول الرجل وبعضه هو واحد و شيء فكانه قال درهم و نصف بين واحد و شيء فاصاب الواحد شيئين فاضرب الشيئين في الواحد والشيء فيكون مالين و شيئين يعدل درهما و نصفا فردهما الي مال واحد وهو ان تاخذ من كل ما معك نصفه فتقول مال و شيء يعدل ثلثة ارباع درهم فقابل به علي نحو ما وصفت لك في صدر الكتاب *

فان قال مال عزلت ثلثه وربعه واربعة دراهم وضربت ما بقي في مثله فعاد المال وزيادة اثني عشر درهما فقياسه انك تاخذ شيئا فتعزل ثلثه وربعه فيبقي خمسة اجزاء من اثني عشر جزءا من شيء فتعزل منها اربعة دراهم فتبقي

فيصير معك اربعة اتساع مال وتسعة دراهم الا اربعة اجذار
يعدل جذرا فزد الاربعة الاجذار علي الجذر فيكون خمسة
اجذار تعدل اربعة اتساع مال وتسعة دراهم فاكمل مالك وهو
ان تضرب الاربعة الاتساع في اثنين وربع فيكون مالا واضرب
تسعة دراهم في اثنين ربع يكون عشرين وربعا ثم اضرب
الخمسة الاجذار في اثنين وربع فيكون احد عشر شيئا وربعا
فيصير معك مال وعشرون درهما وربع يعدل احد عشر
جذرا وربعا فقابل بذلك كنحو ما وصفت لك في تصنيف
الاجذار ان شاء الله *

فان قال مال تضرب ثلثه في ربعه فيعود المال قياسه
ان تضرب ثلث شيء في ربع شيء فيكون نصف سدس
مال يعدل شيئا فالمال يعدل اثني عشر شيئا وهو جذر ماية
واربعة واربعين *

فان قال مال تضرب ثلثه ودرهما في ربعه ودرهمين
فيعود المال و زيادة ثلثة عشر درهما فقياسه ان تضرب ثلث
شيء في ربع شيء فيكون نصف سدس مال و تضرب
درهمين في ثلث شيء فيكون ثلثي جذر ودرهما في ربع
شيء فيكون ربع جذر ودرهمين في درهم درهمان فذلك
نصف سدس مال ودرهمان واحد عشر جزءا من اثني عشر

وكذلك لو قال مال تضرب جذره في اربعة اجذاره فيعود ثلثة امثال المال وزيادة خمسين درهما فقياسه ان تضرب جذرا في اربعة اجذار فيكون اربعة اموال يعدل ثلثة اموال و خمسين درهما فالق ثلثة اموال من الاربعة الاموال يبقي مال واحد يعدل خمسين درهما وهو جذر خمسين مضروب في اربعة اجذار خمسين ايضا فذلك مايتان يكون ثلثة امثال المال وزيادة خمسين درهما   *

فان قال مال تزيد عليه عشرين درهما فيكون مثل اثني عشر جذره فقياسه ان تقول مال و عشرون درهما يعدل اثني عشر جذرا فنتصف الاجذار واضربها في مثلها تكون ستة و ثلثين فانقص منها العشرين الدرهم وخذ جذر ما بقي فانقصه من نصف الاجذار وهو ستة فما بقي وهو جذر المال وهو درهمان والمال اربعة   *

فان قال مال يعزل ثلثه وثلثة دراهم ويضرب ما بقي في مثله فيعود المال فقياسه انك اذا القيت ثلثة وثلثة دراهم بقي ثلثاه الا ثلثة دراهم وهو جذر فاضرب ثلثي شيء الا ثلثة دراهم في مثله فتقول ثلثان في ثلثين اربعة اتساع مال والا ثلثة دراهم في ثلثي شيء جذران والا ثلثة دراهم في ثلثي شيء جذران والا ثلثة دراهم في الا ثلثة دراهم تسعة دراهم

فان قال مال تضربه في اربعة امثاله فيعود ثلث المال الاول فقياسه انك اذا ضربته في اثني عشر مثله عاد المال وهو نصف سدس من ثلث  *

فان قال مال تضربه في جذره فيعود ثلثة امثال المال الاول فقياسه انك اذا ضربت الجذر في ثلث المال عاد المال فتقول هذا مال ثلثه جذره وهو تسعة  *

فان قال مال تضرب اربعة اجذاره في ثلثة اجذاره فيعود المال وزيادة اربعة واربعين درهما فقياسه ان تضرب اربعة اجذار في ثلثة اجذار فيكون اثني عشر مالا يعدل مالا واربعة واربعين درهما فالق من الاثني عشر المال مالا بمال فيبقي احد عشر مالا تعدل اربعة واربعين درهما فاقسمها عليها فيكون اربعة وهو المال  *

فان قال مال تضرب اربعة اجذاره في خمسة اجذاره فيعود مثلي المال وزيادة ستة وثلثين درهما فقياسه انك تضرب اربعة اجذار في خمسة اجذار فيكون عشرين مالا يعدل مالين وستة وثلثين درهما فتلقي من العشرين المال مالين بمالين فيبقي ثمانية عشر مالا يعدل ستة وثلثين درهما فتقسم ستة وثلثين درهما علي ثمانية عشر فيكون القسم اثنين وهو المال  *

فان قال مال ثلثا خمسه مثل سبع جذره فان المال كله يعدل جذرا ونصف سبع جذر فالجذر اربعة عشر جزءا من خمسة عشر جزءا من مال وقياسه ان تضرب ثلثي خمس مال في سبعه ونصف ليتم المال فاضرب ما معك وهو سبع جذر في مثل ذلك فيصير المال يعدل جذرا ونصف سبع جذر ويصير جذره واحدا و نصف سبع فالمال واحد وتسعة وعشرون جزءا من ماية وستة وتسعين من درهم وثلثا خمسه يكون ثلثين جزءا من ماية وستة وتسعين وسبع جذره ايضا ثلثون جزءا من ماية وستة وتسعين *

فان قال مال ثلثة ارباع خمسه مثل اربعة اخماس جذره فقياسه ان تزيد علي ثلثة ارباع خمسه مثل ربعه ليكون الجذر تاما وذلك ثلثة وثلثة ارباع من عشرين فاجعلها ارباعا كليا فيكون خمسة عشر من ثمانين فاقسم الثمانين علي الخمسة عشر فيكون خمسة وثلثا فذلك جذر المال والمال ثمانية وعشرون واربعة اتساع *

وان قال مال تضربه في اربعة امثاله فيكون عشرين فقياسه انك اذا ضربته في مثله كان خمسة وهو جذر خمسة *

فان قال مال تضربه في ثلثه فيكون عشرة فقياسه انك اذا ضربته في مثله كان ثلثين فتقول المال جذر ثلثين *

في مثلها فتكون ماية و مالا الا عشرين شيئا يعدل العشرة الاجذار فقابل بها علي ما قد وصفت لك *

وكذلك لو قال عشرة قسمتها قسمين ثم ضربت احدهما في الاخر ثم قسمت ما اجتمع من الضرب علي فصل ما بين القسمين قبل ان تضرب احدهما في الاخر فخرج خمسة وربعا قياسه ان تاخذ شيئا من العشرة فيبقي عشرة الا شيئا فاضرب احدهما في الاخر فيكون عشرة اجذار الا مالا فهو ما خرج من ضرب احد القسمين في الاخر ثم قسمت ذلك علي فصل ما بين القسمين وهو عشرة الا شيئين فخرج من القسم خمسة وربع ومتي ضربت خمسة وربعا في عشرة الا شيئين خرج لك المال المضروب وهو عشرة اشياء الا مالا فاضرب خمسة وربعا في عشرة الا شيئين يكن اثنين وخمسين درهما ونصفا الا عشرة اجذار ونصفا يعدل عشرة اجذار الا مالا فاجبر الاثنين والخمسين والنصف بالعشرة الاجذار و النصف وزدها علي العشرة الاجذار الا مالا ثم اجبرها بالمال وزد المال علي اثنين وخمسين درهما ونصف فيكون معك عشرون جذرا ونصف جذر يعدل اثنين وخمسين درهما و نصفا و مالا و قابل به علي ما فسرنا في اول الكتاب *

فيبقي ستة اشياء ونصف يعدل درهمين فالشيء الواحد اربعة اجزاء من ثلثة عشر من درهم وباع الستة كل واحد بجزءين من ثلثة عشر من درهم فبلغ ذلك ثمانية وعشرين جزءا من ثلثة عشر من درهم وذلك مثل فضل ما بين الكيلين وهو قفيزان وصرفهما ستة وعشرون جزءا وفضل ما بين السعرين وهو جزءان فذلك ثمانية و عشرون جزءا *

فان قال مالان بينهما درهمان قسمت القليل علي الكثير فاصاب القسم نصف درهم فاجعل احد المالين شيئا والاخر شيئا ودرهمين فلما قسمت شيئا علي شيء ودرهمين خرج القسم نصف درهم وقد علمت انك متي ضربت ما خرج لك من القسم في المقسوم عليه عاد مالك الذي قسمته وهو شيء فقل شيء ودرهمان في النصف الذي هو القسم فيكون نصف شيء ودرهما يعدل شيئا فالقيت نصف شيء بنصف شيء وبقي درهم يعدل نصف شيء فاضعفه يكون الشيء يعدل درهمين والاخر اربعة *

فان قال عشرة قسمتها قسمين فضربت احدهما في عشرة والقسم الاخر في نفسه فاستويا فقياسه ان تضرب شيئا في عشرة فيكون عشرة اشياء ثم تضرب عشرة الا شيئا

٣٥

يعدل احدا وثمانين شيئا فاجبر الماية والمال بالعشرين الشيء وزدها علي الواحد والثمانين فتكون ماية ومالا يعدل ماية جذر وجذرا فنصف الاجذار فيكون خمسين و نصفا واضربها في مثلها فيكون الفين و خمسماية و خمسين وربعا فانقص منها الماية فيبقي الفان واربع ماية وخمسون وربع فخذ جذرها وهو تسعة واربعون ونصف فانقصها من نصف الاجذار وهو خمسون ونصف فيبقي واحد وهو احد القسمين *

فان قال عشرة اقفزة حنطة او شعير بعت كل واحد منهما بسعر ثم جمعت ثمنهما فكان ما اجتمع مثل فصل ما بين السعرين ومثل ما بين الكيلين فخذ ما شيت فانه يجوز فكانك اخذت اربعة وستة فقلت بعت كل واحد من الاربعة بشيء فضربت اربعة في شيء فصار اربعة اشياء وبعت الستة كل واحد بمثل نصف الشيء الذي بعت به الاربعة وان شيئت بثلثه وان شيئت بربعه وما شيئت فانه يجوز فاذا كان بيعك الاخر بنصف شيء فاضرب نصف شيء في ستة فيكون ثلثة اشياء فاجمعها مع الاربعة الاشياء فتكون سبعة اشياء تعدل ما بين الكيلين وهو قفيزان وفصل ما بين السعرين وهو نصف شيء فيكون سبعة اشياء تعدل اثني ونصف شيء فالق نصف شيء من سبعة اشياء

الخمسة الاشياء علي عشرة الا شيئا واخذت نصف ما خرج كان ذلك كقسمك نصف الخمسة الاشياء علي العشرة الا شيئا فاذا اخذت نصف الخمسة الاشياء صار شيئين ونصفا وهو الذي تريد ان تقسمه علي عشرة الا شيئا [يخرج] يعدل خمسين الا خمسة اشياء لانه قال تضم اليه احد القسمين مضروبا في خمسة فيكون ذلك كله خمسين وقد علمت انك متي ضربت ما خرج لك من القسم في المقسوم عليه عاد المال ومالك شيئان ونصف فاضرب عشرة الا شيئا في خمسين الا خمسة اشياء فيكون ذلك خمسماية درهم وخمسة اموال الا ماية شيء يعدل شيئين ونصفا فاردد ذلك الي مال واحد فيكون ذلك ماية درهم ومالا الا عشرين شيئا يعدل نصف شيء فاجبر ذلك الماية وزد العشرين الشيء علي نصف الشيء فيصير معك ماية درهم ومال يعدل عشرين شيئا ونصف شيء فنصف الاشياء واضربها في مثلها وانقص منها الماية وخذ جذر ما بقي وانقصه من نصف الاجذار وهو عشرة وربع فيبقي ثمانية وهو احد القسمين *

---

فان قال عشرة قسمتها قسمين فضربت احد القسمين في نفسه فكان مثل الاخر احد وثمانين مرة فقياس ذلك ان تقول عشرة الا شيئا في مثلها ماية ومال الا عشرين شيئا

الشيء فيكون معك مايه واربعة اموال وسدس مال يعدل احدا واربعين شيئا وثلثي شيء فاردد ذلك الي مال وقد علمت ان المال الواحد من اربعة اموال وسدس هو خمسها وخمس خمسها فخذ من جميع ما معك الخمس وخمس الخمس فيكون معك اربعة وعشرين ومال يعدل عشرة اجذار لان العشرة من احد واربعين شيئا وثلثي شيء خمسها وخمس خمسها فنصف الاجذار وهي خمسة واضربها في مثلها فتكون خمسة وعشرين فانقص منها الاربعة والعشرين التي مع المال يبقي واحد فخذ جذره وهو واحد فانقصه من نصف الاجذار وهي خمسة فبقي اربعة وهو احد القسمين * واعلم بان كل شيئين تقسم هذا علي هذا وهذا علي هذا فانك اذا ضربت الذي يخرج من هذا في الذي يخرج من هذا كان واحدا ابدا *

فان قال عشرة قسمتها قسمين وضربت احد القسمين في خمسة وقسمته علي الاخر ثم القيت نصف ما اجتمع معك وزدته علي المضروب في خمسة فكان خمسين درهما فان قياس ذلك ان تاخذ شيئا من العشرة فتضربه في خمسة فيكون خمسة اشياء مقسومة علي الباقي من العشرة وهو عشرة الا شيئا ماخوذ نصفه ومعلوم انك اذا قسمت

ومال يعدل احد عشر شيئا فنصف الاشياء فتكون خمسة ونصفا فاضربها في مثلها فتكون ثلثين وربعا فانقص منها الثمانية والعشرين التي مع المال فيبقي اثنان وربع فخذ جذر ذلك وهو واحد ونصف فانقصه من نصف الاجذار يبقي اربعة وهو احد القسمين *

فان قال عشرة قسمتها قسمين فقسمت هذا علي هذا وهذا علي هذا فبلغ ذلك درهمين وسدسا * فقياس ذلك انك اذا ضربت كل قسم في نفسه ثم جمعتهما كان مثل احد القسمين اذا ضربت احدهما في الاخر ثم ضربت الذي اجتمع معك من الضرب في الذي بلغ القسم وهو اثنان وسدس فاضرب عشرة الا شيئا في مثلها فتكون ماية ومالا الا عشرين شيئا واضرب شيئا في شيء فيكون مالا فاجمع ذلك فيصير ماية ومالين الا عشرين شيئا يعدل شيئا مضروبا في عشرة الا شيئا وذلك عشرة اشياء الا مالا مضروبا في ما خرج من القسمين وهو اثنان وسدس فيكون ذلك احدا وعشرين شيئا وثلثي شيء الا مالين وسدسا يعدل ماية ومالين الا عشرين شيئا فاجبر ذلك وزد مالين وسدسا علي ماية ومالين الا عشرين شيئا وزد العشرين الشيء الناقصة من الماية والمالين علي الواحد والعشرين الشيء وثلثي

عشرين شيئا فيبقي مابة الا عشرين شيئا يعدل اربعين درهما فاجبر المابة بالعشرين الشيء فزدها علي الاربعين فيكون مابة تعدل عشرين شيئا واربعين درهما فالق الاربعين من المابة فيبقي ستون درهما تعدل عشرين شيئا فالشيء الواحد يعدل ثلثة وهو احد القسمين *

وان قال عشرة قسمتها قسمين فضربت كل قسم في نفسه وجمعتهما وزدت عليهما فصل ما بين القسمين من قبل ان تضربهما فبلغ ذلك اربعة وخمسين درهما فان قياسه ان تضرب عشرة الا شيئا في مثلها فتكون مابة ومالا الا عشرين شيئا وتضرب الشيء الثاني من العشرة في مثله فيكون مالا ثم تجمع ذلك فيكون مابة ومالين الا عشرين شيئا وقال زدت عليهما فصل ما بينهما قبل ان تضربهما فقلت فصل ما بينهما عشرة الا شيئين فجميع ذلك مابة وعشرة ومالان الا اثنين وعشرين شيئا يعدل اربعة وخمسين درهما فاذا جبرت وقابلت قلت مابة وعشرة دراهم ومالان يعدل اربعة وخمسين درهما واثنين وعشرة شيئا فاردد المالين الي مال واحد وهو ان تاخذ نصف ما معك فيكون خمسة وخمسين درهما ومالا يعدل سبعة وعشرين درهما واحد عشر شيئا فالق سبعة وعشرين من خمسة وخمسين فبقي ثمانية وعشرون درهما

## باب المسائل المختلفة *

فان سأل سائل فقال عشرة قسمتها قسمين ثم ضربت احدهما في الاخر فكان واحدا وعشرين درهما فقد علمت ان احد القسمين من العشرة شيء والاخر عشرة الا شيئا فاضرب شيئا في عشرة الا شيئا فيكون عشرة اشياء الا مالا يعدل احدا وعشرين فاجبر العشرة الاشياء بالمال وزده علي الواحد والعشرين فيكون عشرة اشياء تعدل احدا وعشرين درهما ومالا فالق نصف الاجذار فتبقي خمسة فاضربها في مثلها تكن خمسة وعشرين فالق منها الواحد والعشرين التي مع المال فتبقي اربعة فتأخذ جذرها وهو اثنان فانقصه من نصف الاجذار وهي خمسة يبقي ثلثة وذلك احد القسمين وان شئت زدت جذر الاربعة علي نصف الاجذار فيكون سبعة وهو احد القسمين وهذه المسئلة التي تعمل بالزيادة والنقصان *

وان قال عشرة قسمتها قسمين فضربت كل قسم في نفسه ثم القيت الاقل من الاكثر فبقي اربعون قياسه ان تضرب عشرة الا شيئا في مثلها فتكون ماية ومالا الا عشرين شيئا وتضرب شيئا في شيء فيكون مالا فتنقصه من الماية والمال الا

مثلها فتكون خمسة وعشرين فالق منها الواحد والعشرين التي مع المال فيبقي اربعة فخذ جذرها وهو اثنان فانقصه من نصف الاجذار التي هي خمسة فيبقي ثلثة وهو احد القسمين والاخر سبعة فقد اخرجتك هذه المسئلة الي احد الابواب الستة وهو اموال وعدد تعدل جذورا   *:

المسئلة السادسة   * مال ضربت ثلثه في ربعه فعاد المال. وزيادة اربعة وعشرين درهما * فقياسه ان تجعل مالك شيئا ثم تضرب ثلث شيء في ربع شيء فيكون نصف سدس مال يعدل شيئا واربعة وعشرين درهما ثم تضرب نصف سدس مال في اثني عشرحتي تكمل مالك فاضرب الشيء في اثني عشر يكن اثني عشر شيئا واضرب الاربعة والعشرين في اثني عشر فيصير معك مايتان وثمانية وثمانون درهما واثني عشر جذرا يعدل مالا فنصف الاجذار تكون ستة واضربها في مثلها وزدها علي مايتين وثمانية وثمانين فتكون ثلثماية واربعة وعشرين فخذ جذرها وهو ثمانية عشر فزده علي نصف الاجذار وهي ستة فيكون ذلك اربعة وعشرين وهو المال فقد اخرجتك هذه المسئلة الي احد الابواب الستة وهو جذور وعدد تعدل اموالا *

الاجذار واضربها في مثلها تكن اثني عشر وربعا فزدها علي الاعداد وهي مايتان وثمانية وعشرون فتكون مايتين واربعين وربعا فخذ جذرها خمسة عشر ونصفا فانقص منه نصف الاجذار وهو ثلثة ونصف فبقي اثني عشر وهو المال فقد اخرجتك هذه المسئلة الي احد الابواب الستة وهو اموال وجذور تعدل عددا *

والمسئلة الخامسة * عشرة قسمتها قسمين ثم ضربت كل قسم في نفسه وجمعتهما فكانا ثمانية وخمسين درهما * قياسه ان تجعل احد القسمين شيئا والاخر عشرة الا شيئا فاضرب عشرة الا شيئا في مثلها فيكون ماية ومالا الا عشرين شيئا ثم تضرب شيئا في شيء فيكون مالا ثم تجمعهما فيكون ذلك ماية ومالين الا عشرين شيئا يعدل ثمانية وخمسين درهما فاجبر الماية والمالين بالعشرين الشيء الناقصة وزدها علي الثمانية والخمسين فيكون ماية ومالين يعدل ثمانية وخمسين درهما وعشرين شيئا فاردد ذلك الي مال واحد وهو ان تاخذ نصف ما معك فيكون خمسين درهما ومالا يعدل تسعة وعشرين درهما وعشرة اشياء فقابل به وذلك انك تلقي من الخمسين تسعة وعشرين فيبقي احد وعشرون ومال يعدل عشرة اشياء فنصف الاجذار تكون خمسة واضربها في

القسمين شيئا والاخر عشرة الا شيئا ثم تقسم عشرة الا شيئا علي شيء ليكون اربعة وقد علمت انك متي ما ضربت ما خرج لك من القسم في المقسوم عليه عاد المال الذي قسمته والقسم في هذه المسئلة اربعة والمقسوم عليه شيء فاضرب اربعة في شيء فيكون اربعة اشياء تعدل المال الذي قسمته وهو عشرة الا شيئا فاجبر العشرة بالشيء وزده علي الاربعة الاشياء فيكون خمسة اشياء تعدل عشرة فالشيء الواحد اثنان وهو احد القسمين فقد اخرجتك هذه المسئلة الي احد الابواب الستة وهو جذور تعدل عددا  *

والمسئلة الرابعة  *  مال ضربت ثلثه ودرهما في ربعه ودرهم فكان عشرين  *  قياسه ان تضرب ثلث شيء في ربع شيء فيكون نصف سدس مال وتضرب درهما في ثلث درهم فيكون ثلث شيء ودرهما في ربع شيء بربع شيء ودرهما في درهم بدرهم فذلك كله نصف سدس مال وثلث شيء وربع شيء ودرهم يعدل عشرين درهما فالق من العشرين درهما بدرهم فيبقي تسعة عشر درهما تعدل نصف سدس مال وثلث شيء وربع شيء وكمل مالك واكماله ان تضرب كل ما معك في اثني عشر فيصير معك مال وسبعة اجذار يعدل مايتين وثمانية وعشرين درهما فنصف

نفسه والباقي من العشرة اثنان وهو القسم الاخر فقد اخرجتك هذه المسئلة الي احد الابواب الستة وهو اموال تعدل جذورا فاعلم ذلك *

والمسئلة الثانية * عشرة قسمتها قسمين فضربت كل قسم في نفسه ثم ضربت العشرة في نفسها فكان ما اجتمع من ضرب العشرة في نفسها مثل احد القسمين مضروبا في نفسه مرتين وسبعة اتساع مرة او مثل الاخر مضروبا في نفسه ست مرات وربع مرة * فقياس ذلك ان تجعل احد القسمين شيئا والاخر عشرة الا شيئا فتضرب الشيء في نفسه فيكون مالا ثم في اثنين وسبعة اتساع فيكون مالين وسبعة اتساع مال ثم تضرب العشرة في مثلها فيكون ماية تعدل مالين وسبعة اتساع مال فاردده الي مال واحد وهو تسعة اجزاء من خمسة وعشرين جزءا وهو خمس واربعة اخماس الخمس فخذ خمس الماية واربعة اخماس خمسها وهو ستة وثلثون تعدل مالا فخذ جذرها ستة وهو احد القسمين والاخر اربعة لا محالة فقد اخرجتك هذه المسئلة الي احد الابواب الستة وهو اموال تعدل عددا *

والمسئلة الثالثة * عشرة قسمتها قسمين ثم قسمت احدهما على الاخر فخرج القسم اربعة * فقياسه ان تجعل احد

باب المسائل الست * وقد قدمنا قبل ابواب الحساب وجوهه ست مسائل جعلتها امثلة للستة الابواب المتقدمة في صدر كتابي هذا الذي اخبرت ان منها ثلثة لا تنصف فيها الاجذار وذكرت ان حساب الجبر والمقابلة لا بد ان يخرجك الي باب منها ثم اتبعت ذلك من المسائل بما يقرب من الفهم وتحق فيه المؤنة وتسهل فيه الدلالة ان شاء الله تعالي *

فالاولي من الست نحو قولك عشرة قسمتها قسمين فضربت احد القسمين في الاخر ثم ضربت احدهما في نفسه فصار المضروب في نفسه مثل احد القسمين في الاخر اربع مرات * فقياسه ان تجعل احد القسمين شيئًا والاخر عشرة الا شيئًا فتضرب شيئًا في عشرة الا شيئًا فيكون عشرة اشياء الا مالا ثم تضربه في اربعة لقولك اربع مرات فيكون اربعة امثال المضروب من احد القسمين والاخر فيكون ذلك اربعين شيئًا الا اربعة اموال ثم تضرب شيئًا في شيء وهو احد القسمين في نفسه فيكون مالا يعدل اربعين شيئًا الا اربعة اموال فاجبره بالاربعة الاموال وزدها علي المال فيكون اربعين شيئًا يعدل خمسة اموال فالمال الواحد يعدل ثمانية اجذار وهو اربعة وستون جذرها ثمانية وهو احد القسمين المضروب في

مايتين هو جذر ثماني ماية وذلك ما اردنا ان نبين وهذه صورته *

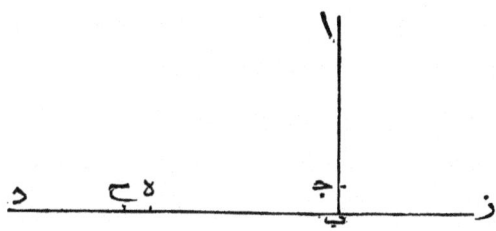

واما ماية ومال الا عشرين جذرا مجموع اليه خمسون وعشرة اجذار الا مالين فلم تستقم له صورة لانه من ثلثة اجناس مختلفة اموال وجذور وعدد وليس معها ما يعادلها فنتصور وقد تمكننا لها صورة لا تحسن فاما اضطرارها باللفظ فبين وذلك انك قد علمت ان معك ماية ومالا الا عشرين جذرا فلما زدت عليها خمسين وعشرة اجذار صارت ماية وخمسين ومالا الا عشرة اجذار لان هذه العشرة الاجذار المزيدة جبرت من العشرين الجذر الناقصة عشرة اجذار فبقيت ماية وخمسون ومال الا عشرة اجذار وقد كان مع الماية مال فلما نقصت من الماية والمالين المستثنيين من الخمسين ذهب مال بمال وبقي عليك مال فصارت ماية وخمسين الا مالا والا عشرة اجذار وذلك ما اردنا ان نبين *

ارايتك في عمل الاضعاف فما بلغ فاقسمه علي اربعة او علي ما اردت ان تقسم عليه واعمل به كما عملت * وكذلك ان اردت ثلثة اجذار تسعة او اكثر او نصف جذر تسعة او اقل او ما كان فعلي هذا القياس فاعمله تصب ان شاء الله تعالي *

وان اردت ان تضرب جذر تسعة في جذر اربعة فاضرب تسعة في اربعة فيكون ستة وثلثين فخذ جذرها وهو ستة وهو جذر تسعة مضروب في جذر اربعة * وكذلك لو اردت ان تضرب جذر خمسة في جذر عشرة فاضرب خمسة في عشرة فجذر ما بلغ هو الشيء الذي تريده * فان اردت ان تضرب جذر ثلث في جذر نصف فاضرب ثلثا في نصف فيكون سدسا فجذر السدس هو جذر الثلث مضروب في جذر النصف * وان اردت ان تضرب جذري تسعة في ثلثة اجذار اربعة فاستخرج جذري تسعة علي ما وصفت لك حتي تعلم جذر اي مال هو وكذلك فافعل بثلثة اجذار الاربعة حتي تعلم جذر اي مال هو ثم اضرب المالين احدهما في الاخر فجذر ما اجتمع لك هو جذري تسعة في ثلثة اجذار اربعة وكذلك كلما زاد من الاجذار او نقص فعلي هذا المثال فاعمل به *

فيكون جذر ما اجتمع مثل نصف جذر ذلك المال * وكذلك ثلثة او اربعة او اقل من ذلك او اكثر بالغا ما بلغ في النقصان والاضعاف * ومثال ذلك اذا اردت ان تضعف جذر تسعة ضربت اثنين في تسعة فيكون ستة وثلثين فخذ جذره يكون ستة وهو ضعف جذر تسعة وكذلك لو اردت ان تضعف جذر تسعة ثلث مرات ضربت ثلثة في ثلثة ثم في تسعة فيكون احد وثمانين فخذ جذره تسعة وذلك جذر تسعة مضاعفا ثلث مرات * فان اردت ان تاخذ نصف جذر تسعة فانك تضرب نصفا في نصف فيكون ربعا ثم تضرب ربعا في تسعة فيكون اثنين وربعا فتاخذ جذرها وهو واحد ونصف وهو نصف جذر تسعة وكذلك ما زاد او نقص من المعلوم والاصم فهذا طريقه *

القسم * وان اردت ان تقسم جذر تسعة علي جذر اربعة فانك تقسم تسعة علي اربعة فيكون اثنين وربعا فجذرها هو ما يصيب الواحد وهو واحد ونصف * وان اردت ان تقسم جذر اربعة علي جذر تسعة فانك تقسم اربعة علي تسعة فيكون اربعة تساع واحد فجذرها ما يصيب الواحد وهو ثلثا واحد * فان اردت ان تقسم جذري تسعة علي جذر اربعة او غيرها من الاموال فاضعف جذر التسعة علي ما

زايدا او ناقصا  مثل الا شيئا في زيادة شيء فالضرب الاخير ناقص ابدا  *  فاعلم ذلك وبالله التوفيق  *

باب الجمع والنقصان  *  اعلم ان جذر مايتين الا عشرة مجموع الي عشرين الا جذرا مايتين فانه عشر سوا  *  وجذر مايتين الا عشرة منقوص من عشرين الا جذر مايتين فهو ثلثون الا جذري مايتين وجذرا مايتين هو جذر ثماني ماية  *  وماية ومال الا عشرين جذرا مجموع اليه خمسون وعشرة اجذار الا مالين فهو ماية وخمسون الا مالا ولا عشرة اجذار  *  وماية ومال الا عشرين جذرا منقوص منه خمسون وعشرة اجذار الا مالين فهو خمسون درهما وثلثة اموال الا ثلثين جذرا  *  وانا مبين لك علة ذلك في صورة تودي الي الباب ان شاء الله تعالي  *  واعلم ان كل جذر مال معلوم او اصم تريد ان تضعفه ومعني اضعافك اياه ان تضربه في اثنين فينبغي ان تضرب اثنين في اثنين ثم في المال فيصير جذر ما اجتمع مثلي [جذر] ذلك المال  *  وان اردت ثلثة امثاله فاضرب ثلثة في ثلثة ثم في المال فيكون جذر ما اجتمع ثلثة امثال جذر ذلك المال الاول وكذلك ما زاد من الاضعاف او نقص فعلي هذا المثال فقسه  *  وان اردت ان تاخذ نصف جذر مال فينبغي ان تضرب نصفا في نصف فيكون ربعا ثم في المال

فيكون عشرة اشياء الا مالا * وان قال عشرة وشيء في شيء الا عشرة قلت شيء في عشرة عشرة اشياء زايدة وشيء في شيء مال زايد و الا عشرة في عشرة ماية درهم ناقصة والا عشرة في شيء بعشرة اشياء ناقصة فتقول مال الا ماية درهم بعد ان قابلت به وذلك ان تضرح عشرة اشياء زايدة بعشرة اشياء ناقصة فيبقي مال الا ماية درهم * وان قال عشرة درهم ونصف شيء في نصف درهم الا خمسة اشياء قلت نصف درهم في عشرة بخمسة دراهم زايدة ونصف درهم في نصف شيء بربع شيء زايد والا خمسة اشياء في عشرة دراهم خمسون جذرا ناقصة فيكون جميع ذلك خمسة دراهم الا تسعة واربعين جذرا وثلثة ارباع جذر ثم تضرب خمسة اجذار ناقصة في نصف جذر زايد فيكون مالين ونصفا ناقصا فذلك خمسة دراهم الا مالين ونصفا والا تسعة واربعين جذرا وثلثة ارباع جذر * فان قال عشرة وشيء في شيء الا عشرة فكانه قال شيء وعشرة في شيء الا عشرة فتقول شيء في شيء مال زايد وعشرة في شيء عشرة اشياء زايدة والا عشرة في شيء عشرة اشياء ناقصة فذهبت الزيادة بالنقصان وبقي المال والا عشرة في عشرة ماية منقوصة من المال فجميع ذلك مال الا ماية درهم * وكل ما كان من الضرب

قلت عشرة في عشرة ماية وعشرة في شيء عشرة اشياء وعشرة في شيء عشرة اشياء ايضا وشيء في شيء مال زايد فيكون ذلك ماية درهم وعشرين شيئا ومالا زايدا * وان قال عشرة الا شيئا في عشرة الا شيئا قلت عشرة في عشرة بماية والا شيئا في عشرة عشرة اشياء ناقصة والا شيئا في عشرة عشرة اشياء ناقصة والا شيئا في الا شيئا بمال زايد فيكون ذلك ماية ومالا الا عشرين شيئا * وكذلك لو انه قال لك درهم الا سدسا في درهم الا سدسا يكون خمسة اسداس في مثلها وهو خمسة وعشرون جزءًا من ستة وثلثين من درهم وهو ثلثان وسدس السدس وقياسه ان تضرب درهما في درهم فيكون درهما والا سدسا في درهم بسدس ناقص والا سدسا في درهم بسدس ناقص فيبقي ثلثان والا سدسا في الا سدسا بسدس السدس زايدا وذلك ثلثان وسدس السدس * فان قال عشرة الا شيئا في عشرة وشيء قلت عشرة في عشرة بماية والا شيئا في عشرة عشرة اشياء ناقصة وشيء في عشرة عشرة اشياء زايدة والا شيئا في شيء مال ناقص فيكون ذلك ماية درهم الا مالا * وان قال عشرة الا شيئا في شيء قلت عشرة في شيء عشرة اشياء والا شيئا في شيء مال ناقص

فالضرب الرابع ناقص * وهو مثل عشرة وواحد في عشرة واثنين فالعشرة في العشرة ماية والواحد في العشرة عشرة زايدة والاثنان في العشرة عشرون زايدة والواحد في الاثنين اثنان زايدان فذلك كله ماية واثنان وثلثون * واذا كانت عشرة الا واحدا في عشرة الا واحدا فالعشرة في العشرة ماية والواحد الناقص في العشرة عشرة ناقصة والواحد الناقص ايضا في العشرة عشرة ناقصة وذلك ثمانون والواحد الناقص في الواحد الناقص واحد زايد فذلك احد وثمانون * واذا كانت عشرة واثنان في عشرة الا واحدا فالعشرة في العشرة ماية والواحد الناقص في العشرة عشرة ناقصة والاثنان الزايدان في العشرة عشرون زايدة فذلك ماية وعشرة والاثنان الزايدان في الواحد المنقوص اثنان ناقصان فذلك كله ماية وثمانية * وانما بينت هذا ليستدل به علي ضرب الاشياء بعضها في بعض اذا كان معها عدد او استثنيت من عدد او استثني منها عدد * فاذا قيل لك عشرة الا شيئا ومعني الشيء الجذري عشرة فاضرب عشرة في عشرة يكون ماية والا شيئا في عشرة يكون عشرة اجذار ناقصة فتقول ماية الا عشرة اشياء * فان قال عشرة وشيء في عشرة فاضرب عشرة في عشرة يكون ماية وشيئا في عشرة بعشرة اشياء زايدة يكون ماية وعشرة اشياء * وان قال عشرة وشيء في مثلها

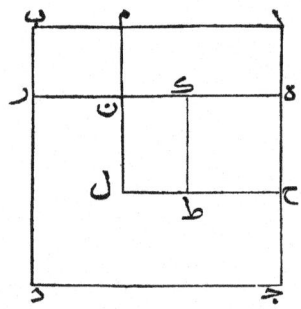

ووجدنا كل ما يعمل به من حساب الجبر والمقابلة لا بد ان يخرجك الي احد الابواب الستة التي وصفت في كتابي هذا وقد اتيت علي تفسيرها فاعرف ذلك *

باب الضرب * وانا مخبرك كيف تضرب الاشياء وهي الجذور بعضها في بعض اذا كانت منفردة او كان معها عدد او كان مستثني منها عدد او كانت مستثناة من عدد وكيف تجمع بعضها الي بعض وكيف تنقص بعضها من بعض * اعلم انه لا بد لكل عدد يضرب في عدد من ان يضاعف احد العددين بعدد ما في الاخر من الاحاد * فاذا كانت عقود ومعها احاد او مستثنيا منها احاد فلا بد من ضربها اربع مرات العقود في العقود والعقود في الاحاد والاحاد في العقود والاحاد في الاحاد * فاذا كانت الاحاد التي مع العقود زايدة جميعا فالضرب الرابع زايد ايضا * واذا كان احدهما زايدا والاخر ناقصا

الاجذار الذي هو واحد ونصف في مثله وهو اثنان وربع ثم زدنا في خط ح‌ج مثل خط اه وهو خط ط‌ل فصار خط ح‌ل مثل خط اح وخط ك‌ن مثل خط ط‌ل وحدث سطح مربع متساوي الاضلاع والزوايا وهو سطح ح‌م وقد تبين لنا ان خط اح مثل خط م‌ل وخط اح مثل خط ح‌ل فبقي خط ح‌ج مثل خط ن‌ر وخط م‌ن مثل خط ط‌ل فنفصل من سطح ه‌ب مثل سطح ك‌ل وقد علمنا ان سطح ا‌ر هو الاربعة الزايدة علي الثلثة الاجذار فصار سطح ان وسطح ك‌ل مثل سطح ا‌ر الذي هو الاربعة العدد فتبين لنا ان سطح ح‌م هو نصف الاجذار الذي هو واحد ونصف في مثله وهو اثنان وربع وزيادة الاربعة التي هي سطح ان وسطح ك‌ل وقد بقي لنا من ضلع المربعة الاولة التي هي سطح اد وهو المال كله نصف الاجذار وهو واحد ونصف وهو خط ح‌ج فاذا زدناه علي خط اح الذي هو جذر سطح ح‌م وهو اثنان ونصف [وزدنا عليه خط ح‌ج الذي هو نصف الثلثة الاجذار وهو واحد ونصف] فبلغ ذلك كله اربعة وهو خط اج وهو جذر المال الذي هو سطح اد وهذه صورته وذلك ما اردنا ان نبين *

الذي هو نصف الاجذار بقي خط اج وهو ثلثة وهو جذر المال الاول * فان زدته علي خط جح الذي هو نصف الاجذار بلغ ذلك سبعة وهو خط رج ويكون جذر مال اكثر من هذا المال اذا زدت عليه واحدا و عشرين صار ذلك مثل عشرة اجذاره وهذا صورته وذلك ما اردنا ان نبين

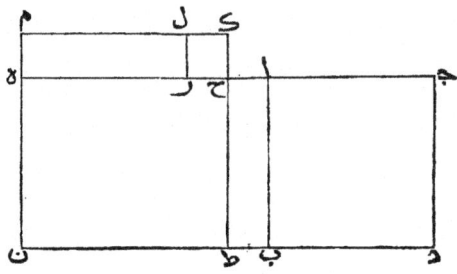

واما ثلثة اجذار واربعة من العدد يعدل مالا فانا نجعل المال سطحا مربعا مجهول الاضلاع متساوي الاضلاع والزوايا وهو سطح اد فهذا السطح كله يجمع الثلثة الاجذار والاربعة التي ذكرناها وكل سطح مربع فان احد اضلاعه في واحد جذره فقطعنا من سطح اد سطح هد فجعلنا احد اضلاعه الذي هو هج الثلثة التي هي عدد الاجذار وهي مثل رد فتبين لنا ان سطح هب هو الاربعة المزيدة علي الاجذار فقطعنا ضلع هج الذي هو ثلثة اجذار بنصفين علي نقطة ح ثم جعلنا منه سطحا مربعا وهو سطح هط وهو ما كان من ضرب نصف

ح فتبين لنا ان خط هح مثل خط ح ج وقد تبين لنا ان خط ح ط مثل خط جد فزدنا علي خط ح ط علي استقامة مثل فصل جح علي ح ط ليتربع السطح فصار خط طك مثل خط كم وحدث سطح مربع متساوي الاضلاع والزوايا وهو سطح م ط وقد كان تبين لنا ان خط طك خمسة واضلاعه مثله فسطحه اذًا خمسة وعشرون وهو ما اجتمع من ضرب نصف الاجذار في مثلها وهو خمسة في خمسة يكون خمسة وعشرين * وقد كان تبين لنا ان سطح هب هو الواحد والعشرون التي زيدت علي المال فقطعنا من سطح هب بخط طك الذي هو احد اضلاع سطح م ط بقي سطح طا * واخذنا من خط كم خط كل وهو مثل خط حك فتبين لنا ان خط طح مثل خط مل وفصل من خط مك خط لك وهو مثل خط كح فصار سطح م ر مثل سطح طا فتبين لنا ان سطح هط مزيدا عليه سطح م ر مثل سطح هب وهو واحد وعشرون وقد كان سطح م ط خمسة وعشرين فلما نقصنا من سطح م ط سطح هط وسطح م ر الذين هما واحد وعشرين بقي لنا سطح صغير وهو سطح رك وهو فصل ما بين خمسة وعشرين وواحد وعشرين وهو اربعة وجذرها خط رح وهو مثل خط ح ا وهو اثنان * فان نقضتيهما من خط ح ج

علي تسعة وثلثين ليتم السطح الاعظم الذي هو سطح ره فبلغ ذلك كله اربعة وستين فاخذنا جذرها وهو ثمانية وهو احد اضلاع السطح الاعظم فاذا نقصنا منه مثل ما زدنا عليه وهو خمسة بقي ثلثة وهو ضلع سطح اب الذي هو المال وهو جذره والمال تسعة وهذه صورته

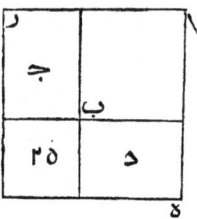

واما مال واحد وعشرون درهما يعدل عشرة اجذاره فانا نجعل المال سطحا مربعا مجهول الاضلاع وهو سطح اد ثم نضم اليه سطحا متوازي الاضلاع عرضه مثل احد اضلاع سطح اد وهو ضلع هن والسطح هب فصار طول السطحين جميعا ضلع جه وقد علمنا ان طوله عشرة من العدد لان كل سطح مربع متساوي الاضلاع والزوايا فان احد اضلاعه مضروبا في واحد جذر ذلك السطح وفي اثنين جذراه فلما قال مال واحد وعشرون يعدل عشرة اجذاره علمنا ان طول ضلع هج عشرة اعداد لان ضلع جد جذر المال فقسمنا ضلع جه بنصفين علي نقطة

ليتم لنا بناء السطح الاعظم بما نقص من زواياه الاربع لان كل عدد يضرب ربعه في مثله ثم في اربعة يكون مثل ضرب نصفه في مثله فاستغنينا بضرب نصف الاجذار في مثلها عن الربع في مثله ثم في اربعة وهذا صورته.

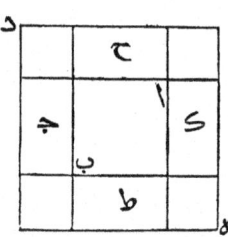

وله ايضا صورة اخرى تودي الي هذا وهي سطح اب وهو المال فاردنا ان نزيد عليه مثل عشرة اجذاره فنصفنا العشرة فصارت خمسة فصيرناها سطحين على جنبتي سطح اب وهما سطحا ج د فصار طول كل سطح منهما خمسة اذرع وهو نصف العشرة الاجذار وعرضه مثل ضلع سطح اب فبقيت لنا مربعة من زوايا سطح اب وهي خمسة في خمسة وهي نصف العشرة الاجذار التي زدناها على جنبتي السطح الاول فعلمنا ان السطح الاول هو المال وان السطحين الذين على جنبتيه هما عشرة اجذار فذلك كله تسعة وثلثون وبقي الي تمام السطح الاعظم مربعة خمسة في خمسة فذلك خمسة وعشرون فزدناها

فهو جذره وكل ضلع من اضلاعه اذا ضربته في عدد من الاعداد فما بلغت الاعداد فهي اعداد جذور * كل جذر مثل جذر ذلك السطح فلما قيل ان مع المال عشرة اجذاره اخذنا ربع العشرة وهو اثنان و نصف وصيرنا كل ربع منها مع ضلع من اضلاع السطح فصار مع السطح الاول الذي هو سطح ا ب وعرضه اثنان وهي نصف وهي سطوح ج ط ك ج فحدث سطح متساوي الاضلاع مجهول ايضا ناقص في زواياه الاربع في كل زاوية من النقصان اثنان و نصف في اثنين و نصف فصار الذي يحتاج اليه من الزيادة حتي يتربع السطح اثنان و نصف في مثله اربع مرات و مبلغ ذلك جميعه خمسة وعشرون * وقد علمنا ان السطح الاول الذي هو سطح المال والاربعة السطوح التي حوله وهي عشرة اجذاره تسعة وثلثون من العدد * فاذا زدنا عليها الخمسة و العشرين التي هي المربعات الاربع التي هي علي زوايا سطح ا ب تم تربيع السطح الاعظم وهو سطح د ه وقد علمنا ان ذلك كله اربعة وستون واحد اضلاعه جذره وهو ثمانية فاذا نقصنا من الثمانية مثل ربع العشرة مرتين من طرفي ضلع السطح الاعظم الذي هو سطح د ه وهو خمسة بقي من ضلعه ثلثة وهو جذر ذلك المال * وانما نصفنا العشرة الاجذار وضربناها في مثلها وزدناها علي العدد الذي هو تسعة وثلثون

مثل نصف الاجذار سوا لا زيادة ولا نقصان وكل ما اتاك من مالين او اكثر او اقل فاردده الي مال واحد كنحو ما بينت لك في الباب الاول *

واما الجذور والعدد التي تعدل الاموال فنحو قولك ثلثة اجذار واربعة من العدد يعدل مالا فقياسه ان تنصف الاجذار فتكون واحدا ونصفا فاضربها في مثلها فتكون اثنين و ربعا فزدها علي الاربعة فتكون ستة و ربعا فخذ جذرها وهو اثنان و نصف فزده علي نصف الاجذار وهو واحد و نصف فيكون اربعة وهو جذر المال والمال ستة عشر وكل ما كان اكثر من مال او اقل فاردده الي مال واحد *

فهذه الستة الضروب التي ذكرتها في صدر كتابي هذا وقد اتيت علي تفسيرها واخبرت ان منها ثلثة ضروب لا تنصف فيها الاجذار وقد بينت قياسها واضطرارها * فاما ما يحتاج فيه الي تنصيف الاجذار من الثلثة الابواب الباقية فقد وصفته بابواب صحيحة و صيرت لكل باب منها صورة يستدل بها علي العلة في التنصيف *

فاما علة مال و عشرة اجذار يعدل تسعة وثلثين درهما فصورة ذلك سطح مربع مجهول الاضلاع وهو المال الذي تريد ان تعرفه و تعرف جذره وهو سطح اب وكل ضلع من اضلاعه

عشر ونصفه ثمانية * وكذلك فافعل بجميع ما جاءك من الاموال والجذور وما عادلها من العدد يصب ان شاء الله *

واما الاموال والعدد التي تعدل الجذور فنحو قولك مال واحد وعشرون درهما من العدد يعدل عشرة اجذاره ومعناه اي مال اذا زدت عليه واحدا وعشرين درهما كان ما اجتمع مثل عشرة اجذار ذلك المال * فقياسه ان تنصف الاجذار فيكون خمسة فاضربها في مثلها يكون خمسة وعشرين فانقص منها الواحد والعشرين التي ذكر انها مع المال فيبقي اربعة فخذ جذرها وهو اثنان فانقصه من نصف الاجذار وهي خمسة فيبقي ثلثة وهو جذر المال الذي تريده والمال تسعة وان شيت فزد الجذر علي نصف الاجذار فيكون سبعة وهو جذر المال الذي تريده والمال تسعة واربعون * فاذا وردت عليك مسئلة تخرجك الي هذا الباب فامتحن صوابها بالزيادة فان لم تكن فهي بالنقصان لا محالة وهذا الباب يعمل بالزيادة والنقصان جميعا وليس ذلك في غيره من الابواب الثلثة التي تحتاج فيها الي تنصيف الاجذار *

واعلم انك اذا نصفت الاجذار في هذا الباب وضربتها في مثلها فكان مبلغ ذلك اقل من الدراهم التي مع المال فالمسئلة مستحيلة وان كان مثل الدراهم بعينها فجذر المال

اذا جمعا وزيد عليهما مثل عشرة اجذار احدهما بلغ ذلك ثمانية واربعين درهما فينبغي ان ترد المالين الي مال واحد وقد علمت ان مالا من مالين نصفهما فاردد كل شيء في المسئلة الي نصفه فكانه قال نصف مال وخمسة اجذار يعدل اربعة وعشرين درهما ومعناه اي مال اذا زدت عليه خمسة اجذاره بلغ ذلك اربعة وعشرين فنصّف الاجذار فتكون اثنين ونصفا فاضربها في مثلها فتكون ستة وربعا فردها علي الاربعة والعشرين فتكون ثلثين درهما وربعا فخذ جذرها وهو خمسة ونصف فانقص منها نصف الاجذار وهو اثنان و نصف تبقي ثلثة وهو جذر المال والمال تسعة * وكذلك لو قال نصف مال وخمسة اجذار يعدل ثمانية و عشرين درهما فمعني ذلك اي مال اذا زدت علي نصفه مثل خمسة اجذاره بلغ ذلك ثمانية و عشرين درهما فتريد ان تكمل مالك حتي يبلغ مالا تاما وهو ان تضعفه فاضعفه واضعف كلما معك مما يعادله فيكون مالا وعشرة اجذار يعدل ستة وخمسين درهما فنصّف الاجذار تكون خمسة فاضربها في مثلها تكون خمسة وعشرين فزدها علي الستة والخمسين تكون احدا وثمانين فخذ جذرها وهو تسعة فانقص منه نصف الاجذار وهو خمسة فيبقي اربعة وهو جذر المال الذي اردته والمال ستة

اربعة اجذار تعدل عشرين والجذر الواحد يعدل خمسة والمال الذي يكون منه خمسة وعشرون * وكقولك نصف جذر يعدل عشرة فالجذر يعدل عشرين والمال الذي يكون منه اربعماية *

ووجدت هذه الضروب الثلثة التي هي الجذور والاموال والعدد يقترن فيكون منها ثلثة اجناس مقترنة وهي اموال وجذور تعدل عددا * واموال وعدد تعدل جذورا وجذور وعدد تعدل اموالا *

فاما الاموال والجذور التي تعدل العدد فمثل قولك مال وعشرة اجذاره يعدل تسعة وثلثين درهما ومعناه اي مال انا زدت عليه مثل عشرة اجذار بلغ ذلك كله تسعة وثلثين * فقياسه ان تنصف الاجذار وهي في هذه المسئله خمسة فتضربها في مثلها فيكون خمسة وعشرين فتزيدها علي التسعة والثلثين فيكون اربعة وستين فتاخذ جذره وهو ثمانية فتنقص منه نصف الاجذار وهو خمسة فيبقي ثلثة وهو جذر المال الذي تريد والمال تسعة * وكذلك لو ذكر مالين او ثلثة او اقل او اكثر فاردده الي مال واحد واردد ما كان معه من الاجذار والعدد الي مثل ما رددت اليه المال * وهو نحو قولك مالين وعشرة اجذار يعدل ثمانية واربعين درهما ومعناه اي مالين

فاما الاموال التي تعدل الجذور فمثل قولك مال يعدل خمسة اجذاره فجذر المال خمسة والمال خمسة وعشرون وهو مثل خمسة اجذاره * وكقولك ثلث مال يعدل اربعة اجذار فالمال كله يعدل اثني عشر جذرا وهو ماية واربعة واربعون وجذره اثني عشر * ومثل قولك خمسة اموال تعدل عشرة اجذار فالمال الواحد يعدل جذرين وجذر المال اثنان والمال اربعة * وكذلك ما كثر من الاموال او قل يرد الي مال واحد وكذلك يفعل بما عادلها من الاجذار يرد الي مثل ما يرد اليه المال *

واما الاموال التي تعدل العدد فمثل قولك مال يعدل تسعة فهو المال وجذره ثلثة * وكقولك خمسة اموال تعدل ثمانين فالمال الواحد خمس الثمانين وهو ستة عشر * وكقولك نصف مال يعدل ثمانية عشر فالمال يعدل ستة وثلثين وجذره ستة * وكذلك جميع الاموال زايدها وناقصها ترد الي مال واحد وان كانت اقل من مال زيد عليها حتي تكمل مالا تاما و كذلك تفعل بما عادلها من الاعداد *

واما الجذور التي تعدل عددا فكقولك جذر يعدل ثلثة من العدد فالجذر ثلثة والمال الذي يكون منه تسعة * وكقولك

واني لما نظرت فيما يحتاج اليه الناس من الحساب وجدت جميع ذلك عددا ووجدت جميع الاعداد انما تركبت من الواحد والواحد داخل في جميع الاعداد * ووجدت جميع ما يلفظ به من الاعداد ما جاوز الواحد الي العشرة يخرج مخرج الواحد ثم تثني العشرة و تثلث كما فعل بالواحد فيكون منها العشرون والثلثون الي تمام الماية ثم تثني الماية و تثلث كما فعل بالواحد وبالعشرة الي الالف ثم كذلك يردد الالف عند كل عقد الي غاية المدرك من العدد *

ووجدت الاعداد التي يحتاج اليها في حساب الجبر والمقابلة علي ثلثة ضروب وهي جذور و اموال وعدد مفرد لا ينسب الي جذر ولا الي مال * فالجذر منها كل شيء مضروب في نفسه من الواحد وما فوقه من الاعداد وما دونه من الكسور * والمال كلما اجتمع من الجذر المضروب في نفسه * والعدد المفرد كل ملفوظ به من العدد بلا نسبة الي جذر ولا الي مال * فمن هذه الضروب الثلثة ما يعدل بعضهم بعضا وهو كقولك اموال تعدل جذورا * واموال تعدل عددا * وجذور تعدل عددا *

اما رجل سبق الي ما لم يكن مستخرجا قبله فورثه من بعده واما رجل شرح مما ابقا الاولون ما كان مستغلقا فاوضح طريقه وسهل مسلكه وقرب ماخذه واما رجل وجد في بعض الكتب خللا فلّم شعثه واقام اوده واحسن الظن بصاحبه غير زار عليه ولا مفتخر من ذلك بفعل نفسه *

وقد شجعني ما فضل الله به الامام المامون امير المومنين مع الخلافة التي جاز له ارثها واكرمه بلباسها وحلاه بزينتها من الرغبة في الادب وتقريب اهله وادناءهم وبسط كنفه لهم وبمعونته اياهم علي ايضاح ما كان مستبهما وتسهيل ما كان مستوعرا علي ان الفت من حساب الجبر والمقابلة كتابا مختصرا حاصرا للطيف الحساب وجليله لما يلزم الناس من الحاجة اليه في موارثتهم ووصاياهم وفي مقاسمتهم واحكامهم وتجاراتهم وفي جميع ما يتعاملون به بينهم من مساحة الارضين وكري الانهار والهندسة وغير ذلك من وجوهه وفنونه مقدما لحسن النية فيه وراجيا لان يبذله اهل الادب بفضل ما استودعوا من نعم الله تعالي وجليل اياه وجميل بلاية عندهم منزلته وبالله توفيقي في هذا وفي غيره عليه توكلت وهو رب العرش العظيم وصلي الله علي جميع الانبياء والمرسلين *

بسم الله الرحمن الرحيم

هذا كتاب وضعه محمد بن موسى الخوارزمي افتتحه بان قال الحمد لله على نعمه بما هو اهله من محامده التي بداء ما افترض منها على من يعبده من خلقه نقع اسم الشكر ونستوجب المزيد ونومن من الغير اقرارا بربوبيته وتذللا لعزته وخشوعا لعظمته بعث محمدا صلى الله عليه وعلى آله وسلم بالنبوة على حين فترة من الرسل وتنكر من الحق ودروس من الهدى فبصر به من العمى واستنقذ به من الهلكة وكثر به بعد القلة والف به بعد الشتات تبارك الله ربنا وتعلى جده وتقدست اسماؤه ولا اله غيره وصلى الله على محمد النبي وآله وسلم *

ولم تزل العلماء في الازمنة الخالية والامم الماضية يكتبون الكتب مما يصنفون من صنوف العلم ووجوه الحكمة نظرا لمن بعدهم واحتسابا للاجر بقدر الطاقة ورجاء ان يلحقهم من اجر ذلك وذخره وذكره ويبغي لهم من لسان الصدق ما يصغر في جنبة كثير مما كانوا يتكلفونه من المؤونة ويحملونه على انفسهم من المشقة في كشف اسرار العلم وغامضه *

الكتاب المختصر

في حساب الجبر و المقابلة

تصنيف

الشيخ الاجل ابي عبد الله محمد بن موسي

الخوارزمي

طبع في مدينة لندن
سنة ١٨٣٠ المسيحية

الكتاب المختصر

في

حساب الجبر و المقابلة